Driving to Net 0

Stories of Hope for a Carbon-Free Future

For our children and grandchildren

Scientists say we need to reduce our use of fossil fuels by 80% or more to avert the most dangerous effects of global warming. But is such a drastic cut possible without totally disrupting our lifestyle? The short answer is yes, as the contributors of this book show. They imagined a future where each home powers not only itself, but the cars surrounding it. They do this with no fumes, no soot, and no emissions, from a fuel source that will never run out or jump in price. They show this future is possible now as **they drive to net 0**.

Chapters

Acknowledgements

Pulling this book together took the help of many people. I especially want to recognize Debbie Coleman, Ron Freund and Jim Hardy, who edited--a thankless job that was much needed and appreciated.

I also would like to thank Ed Montgomery and Steven Barry, who helped with initial scope and layout of the book, and my wife Brenda, who put up with the 18+ month process of bringing this book to market.

About the author:

Dave Hrivnak received a BS in engineering from Va. Tech and an MBA from East Tennessee State University. In his spare time Dave has been actively finding ways to live sustainably. He loves to push the boundaries of energy efficiency on projects such as: designing and building two passive solar homes, one earth bermed, built an all-electric Jeep & Miata, and modified their Prius and Chevy Avalanche into plug-in hybrid cars. He and his wife Brenda now power their home and vehicles from rooftop solar.

Dave is active in the community as a Boy Scout leader, Habitat for Humanity volunteer, and a mentor with the First Robotics program. Dave leads the energy task force at his church to reduce costs and help the environment. Their home has been featured several times on the National Solar Tour of Homes, and he has coordinated five National Drive Electric events in the Tri-Cities.

Imagine a future where each home powers not only itself, but the cars surrounding it. Imagine doing this with no fumes, no soot, and no emissions, from a fuel source that will never run out or jump in price. That future is possible now as many households have discovered; **they are driving to net 0**.

The Boy Scouts have a saying, "Always leave a place better than you found it." Not only is that good advice for a back-country campsite, but great advice for the Earth as well. In the 1980's we only had the vaguest understanding that there was a serious downside to the incredible benefits of fossil fuels. But now the evidence is clear. Fortunately, we are also on the cusp of an exciting wave of clean energy innovations. Solar and other renewables are rapidly gaining ground; electric cars are far cleaner[1] and perform better than their gasoline counterparts; advances in energy efficiency and lighting allow us to do more with much less energy. So, we now have an option to leave this world better than we found it.

This is encouraging, as climate scientists say we need to cut carbon dioxide (CO_2) emissions by 80 percent to avert dangerous levels of warming.[2] Such deep cuts sound daunting, if not impossible. So, we set out to find what can be done today and are pleased to discover it is possible to drastically cut one's own emissions, and costs, with today's technology. Along our journey, we connected with others who are making their own sustainable journeys. What follows in this book are their stories--households that have made great cuts in net energy emissions. The following chapters are not meant as a "one size fits all" solution to our current lifestyles, but to show several possible paths. The reasons for this journey are varied, as not all have done this for environmental reasons. For some, the primary driver is to save money, while others want energy independence. But all these households have made at least a 75% reduction in their emissions or are 75% below the "average" family. And several have gone well beyond becoming carbon negative, showing, in a very practical sense, that an 80+% reduction is feasible today.

The various authors are spread across the USA and Canada, covering a wide range of climates and geographies.

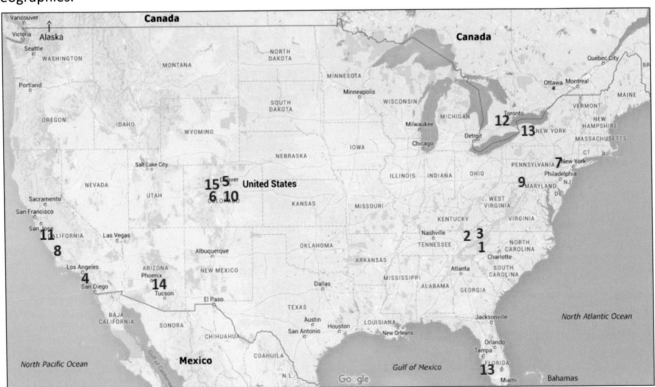

Figure 1 - Chapter authors cover a wide range of climates from coast to coast. The numbers indicate the chapter of the authors.

We hope you find inspiration in these stories and learn how there is an exciting, cleaner, greener future awaiting you.

How much CO_2 does the average family generate?

This book focuses on households for the simple reason that household energy use is a big slice of our energy pie, accounting for about 40 percent of carbon emissions. In the United States and Canada, the average household generates nearly 42,000 pounds of CO_2 from energy use every year. This does not count CO_2 from purchases and food production. This personal/home energy category is the largest single category of emissions in the U.S. Therefore, households need to be part of the solution.

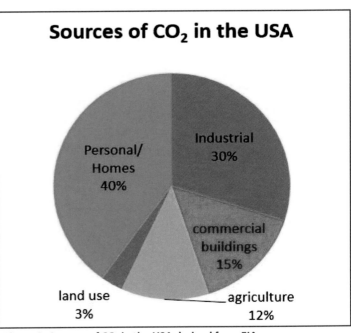

Figure 2- Sources of CO_2 in the USA derived from EIA.

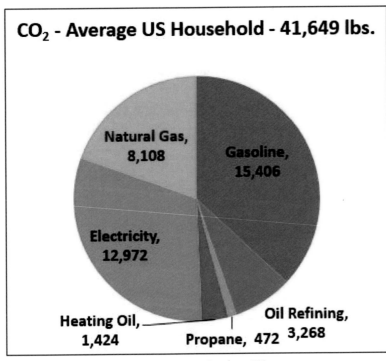

Figure 3 - Average Household CO_2 generation from EIA.

The largest CO_2 source for the typical family is transportation, generating over 18,000 pounds of CO_2 each year, through the burning of gasoline or diesel and its direct refining.[3] If cars were to plop out their emissions visibly, like horses, we would see a pound of waste every mile.[4]

The second largest source of CO_2 for most homeowners is electricity, followed by natural gas and heating oil. Propane brings up the rear. But remember, there is no "average" household, as different homes use different fuel sources for winter heating, and we live in many different climates.

Additionally, emissions vary greatly from country to country. United States' emissions are double Germany's and triple those of Switzerland.[5] What is shocking about this statistic is that both those countries have a higher standard of living than the U.S.[6] Reducing energy consumption does not equate to a lower standard of living.

If you want to calculate your own CO_2 footprint, which we strongly suggest you do, use the numbers on the following page to get a rough guide of your emissions. Review last year's bills and enter your annual numbers into the following conversion table. The one number with large variation is electricity generation. In an area with primarily coal generation, like the Midwest, CO_2 emissions can be as high as 2 lbs./kWh. In an area with a lot of nuclear, hydro, and renewables such as Vermont, the number can be below .1.[7] Most utilities publish a sustainability report reporting their CO_2 emissions/kWh. You can use your local utility's sustainability report to find your exact factor for electricity.

Energy Consumption				Lbs. of CO_2
Gallons of gasoline burned annually	_____	* 23.4[8]	=	_____
Gallons of diesel burned annually	_____	* 26.0[9]	=	_____
kWh of electricity used annually	_____	* 1.15[10]	=	_____
Natural Gas Mcf used annually	_____	* 115[11]	=	_____
Heating Oil gallons annually	_____	* 22.4[12]	=	_____
Propane gallons annually	_____	* 12.7[13]	=	_____
		Total		_____

How to read this book?

This book is more akin to a book of short stories than a novel. Each chapter can stand on its own and is written by a different author meeting a different set of needs. If for some reason a chapter is not resonating with you, feel free to move to the next. Hopefully, you will find several authors and stories you can identify with and begin to put their practices into action. The book features homes from 700 to 10,000 sq. ft., from bicycle commuters to electric vehicle evangelists, from extensive remodels to new construction using the latest technology, and from meat eaters to vegans. Some stories feature full families, and others are empty nesters or couples just starting out. But all have managed to make drastic cuts in their energy emissions. Toward the beginning of each chapter there is a "Quick Facts" box to highlight the emission savings, basic home facts, technologies used, and cost savings. Use this as a guide to how the chapter may align with your lifestyle.

Quick Facts:
    Net Energy Emissions –
    Home – size and location
    Technologies Employed –
    Annual Net Energy Costs –

The summary chapter attempts to tie the stories together, and we conclude with a glossary of terms that may be new to you.

Chapter 1 - An Unapologetic Car Junkie   by Dave Erb

I love cars. Actually, I love vehicles: skateboards to spacecraft, biplanes to bass boats, and everything in between. But I *really* love cars. Some of my first and fondest memories are associated with the ambitious road trips my family took every year. I learned early that cars granted access to the world, or at least to the continent. In 1969, the year I turned 13, my mother hauled four kids and assorted friends to Key West, FL in the spring, then to Anchorage, AK in the summer. We camped right beside the Al-Can Highway, tents pitched next to a white-over-green VW bus, 23 windows and a canvas sunroof, jeep cans and spare tires in a roof rack over the driver and front passenger. If there was a bridge to Hawaii, we'd have driven there, too.

Figure 4 - Racing Clare Bell's electric Porsche in EVTC's Street Stock class at Phoenix in '98.

Quick Facts:
  Net Energy Emissions – 1250 lbs. - 97% less than the typical American family
  Home – 1950's Traditional home, 1,400 square feet in Western North Carolina, USA
  Technologies Employed – Ground source HP, heat pump water heater, solar PV, electric car
  Annual Energy Costs – $285, down 94%

Truth be told, the access and the experiences and the vehicles had become completely intermingled long before 1969. And access, though certainly desirable and of great interest, was just one facet of the enjoyment of automobiles, and not entirely necessary, at that. Cars had already acquired meaning and beauty, in and of themselves, as cultural artifacts, even in the total absence of practical transportation value. The watershed year was 1963.

I had been vaguely aware of auto racing. My mother used to tell stories of family picnics in her childhood backyard, listening to the Indianapolis 500 on the radio, back when 500 miles meant an all-day affair. But racing had no real presence in my life until 1963, when I saw an advertisement in a magazine. I don't recall the product, but I still remember the layout of that ad. A picture of 1963 Indy winner Parnelli Jones was featured near the top, with smaller pictures of Jim Clark (2nd) and A.J. Foyt (3rd) side by side below him. But the pictures weren't really of the drivers. They were pre-race photos of the drivers in their cars. And two of those photos might as well have been invisible. Jim Clark's mid-engine Lotus was the shining green and gold apogee of technology, as sleek and modern as NASA's rockets, just then in transition from Mercury to Gemini. The traditional, front-engine roadsters Jones and Foyt drove could have been oxcarts, for all I cared. The existential truth of art had revealed itself to a seven-year-old.

I eventually went to mechanical engineering school, with the dream of emulating race driver Mark Donohue, an engineer who is still the only sports figure I've ever considered a role model. But Donohue died, after walking away from the crash that later killed him. And money issues temporarily halted my schooling. When I went back, it was in aerospace engineering. I earned bachelor's and master's degrees in aerospace and ocean engineering from Virginia Tech, doing my master's thesis research on composite materials for NASA Langley. I celebrated finishing grad school with an aerospace themed vacation to Florida, where my younger brother was graduating from flight school in Daytona Beach, and NASA was launching the very first space shuttle mission from Cape Canaveral.

But the call of the wheeled proved too strong. My first job after graduate school was as a test and development engineer for a heavy truck manufacturer, working mainly on fuel economy improvement. Next up came fuel economy and emissions work for a transit bus maker. There, in 1986, I had my first opportunity to work with HEVs. Eventually, I ended up in academia, advising a series of student projects that led to a specialty in EVs, especially

Figure 5 - The University of Idaho 1999 race entry with AC Propulsion drivetrain & me driving.

HEVs. I've raced EVs and ICVs on road courses, dirt and paved ovals, drag strips, and autocross courses. I've held calculators, wrenches, steering wheels, and checkbooks in pursuit of higher speeds and lower times. After decades spent studying reams of data, objective and subjective, on virtually all the likely candidates for transportation energy, it's clear to me that electrified powertrains are the inevitable future of land travel.

NASCAR is Not the Enemy

Fascinating though they may be, cars aren't the only interesting, important things in the world. Those childhood trips took us to a lot of wonder-filled places; not all unique, but many magnificent, beautiful, awe-inspiring. Some of my finest moments have been spent outdoors: hiking, fishing, hunting, skiing, cycling, gardening. "Sacred" is an appropriate word for more than a few places I've experienced, including some so seemingly mundane that they'd barely register on most people's radar. Your own back yard is a place of infinite inspiration, if you're aware enough to notice.

Any thinking outdoorsman recognizes the importance of the natural world, and any thinking gardener acknowledges the majesty (and, more importantly, the mystery) of the divine. Both should proudly identify themselves as "environmentalists," a word that an army of lying shills have turned into a slur in the eyes of many people who are smart enough to know better. Sleazy professional sophists attempt to rationalize the sins of their patrons by fabricating "jobs vs. environment" narratives. People of good will need to recognize these sociopaths for who they are and call them out on it. The economy is a wholly-owned subsidiary of the environment, not a separate entity or, God forbid, the parent company. People who know "the price of everything, and the value of nothing" aren't just foolhardy, they're dangerous. So, yes, I'm an environmentalist, and damn proud of it.

Many of my fellow environmentalists have a hard time reconciling environmentalism with automotive enthusiasm. A tiny fraction even argue that cars are evil and, therefore, drivers are, too. A further tiny fraction of those actually practice what they preach, or believe they do, anyway. I once unsubscribed from the transportation listserv of a major environmental organization after a California-based EV advocate was repeatedly abused by some of the other participants. One of the worst offenders was a New Englander who claimed never to drive. What the self-righteous Puritan conveniently ignored, however, was that cars aren't the only vehicles that consume energy and emit pollution. The ferry (yes, ferry!) he rode to work used far more energy per passenger mile than a Nissan Leaf, grossly outweighing any savings he might have realized by riding his bicycle to the terminal. And ferries burn bunker oil, the filthiest transportation fuel around, now that coal-burning vehicles have been relegated to museums. Meanwhile, the heathen Californian charged his Nissan Leaf and Zero electric motorcycle from PV panels on his home's roof.

Such clueless extremists provide ammunition to the fossil fuel fans, who argue that any environmentalist who uses motorized transportation is a hypocrite, and anyone who doesn't is some kind of weirdo. (How convenient, to be able to unilaterally dismiss the authority of everyone who might disagree with you.) Fortunately, however, the ascetic zealots are unusual, and easily recognized as the outliers they are. They're a relatively minor cross to bear.

That's not to say, however, that anti-car sentiment is rare in environmental circles. It's just that most of us have developed more subtle ways of pointing out other people's faults, while ignoring our own. So, people who commute by car every day will direct their scorn, not at cars and drivers, per se, but at car culture. Cars aren't evil, they rationalize, but liking cars is. Racing, in particular, is targeted as wasteful and unnecessary.

It's true that a NASCAR stock car, which gets between two and five miles per gallon of premium racing gas (depending on the track where it's racing), consumes more energy than a kayak, a bicycle, or a pair of hiking boots. But arguing from that perspective misses the point. In the grand scheme of sporting energy consumption, the fuel burned by race cars is trivial.

At two miles per gallon, the 40 cars which start the Daytona 500 consume 10,000 gallons of gas, should they all miraculously finish the entire event. Now, consider a major college football game, with 80,000 fans in the stands. If every one of those fans arrived in a four-person Prius carpool (what are the odds?), Daytona's race fuel would support an average round trip of 25 miles for every vehicle in the parking lot. How many alumni live within thirteen miles of their alma mater? And, what about the opposing team's fans? Duke and Carolina may be crosstown rivals, but Tennessee and Florida sure aren't.

A kayak may not burn any fuel, but few people have a whitewater river in their backyards, and even those who do generally want to paddle more than a few feet before portaging back upstream. A bicyclist can tour the world without burning a single drop of gas, but there are an awful lot of bike (and kayak, and ski) racks on the cars at my Sierra Club meetings. I have yet to hear anyone at those meetings suggest that we should outlaw driving to our outdoor playgrounds, or that football fans should be required to hike to the stadium. Suggesting that we ban racing is just as ridiculous.

All of this would be pointless academic discussion, if not for the fact that there are consequences to making enemies. Strident purists, even when they aren't espousing total BS, can alienate a lot of potential allies. Meeting people where they are, even when they don't look like allies, can win a lot of friends, who can turn into allies over time. And, Lord knows, we need allies. No one person is solely responsible for the environmental challenges we face; all of us are contributing to the problem. Which means that we all, singly and together, need to start doing less damage if we're to clean up our mess. Racing, racers, and auto enthusiasts can help.

The Flea that Bites the Tail that Wags the Dog

It's no revelation that Americans drive a lot. In 2014, approximately 214 million licensed drivers registered 263 million vehicles in the U.S., over six cars for every five drivers. We drove them about three trillion miles. Though very few of those miles were driven purely for enjoyment, most all of those drivers have a personal connection to their cars. Cars aren't frigidly rational calculations on a financial spreadsheet. They're red hot, emotionally charged fashion statements, even for people who think buying refrigerator white Toyota Camrys proves they're above the fray. Those emotional connections allow people to see their energy consumption, to feel it at a gut level that LED lights and better insulation simply can't touch. People who don't bat an eye when dropping tens of thousands to buy a car will become livid when gas prices go up a quarter a gallon, but they still buy the gas. And that presents opportunities to advocates who are alert enough to seize them.

American auto enthusiasts, perhaps five million strong, represent a fraction of American drivers. Active racers, perhaps 100,000 in all, represent an even smaller fraction of enthusiasts, and fewer than a thousand race at levels that are visible to the public. But enthusiasts are society's opinion leaders for cars, the people non-enthusiasts consult when purchasing vehicles. And racers command inordinate sway among enthusiasts, despite the fact that racing is far removed from anything that anyone who's not psychotic would ever do on the street.

From 1995 through 2004, I created and taught a three-day professional development seminar on "Design of Hybrid Electric Vehicles" for the Society of Automotive Engineers (SAE). Though I already had extensive experience developing both HEVs and BEVs when I started teaching the course, my credibility with my audience rose noticeably after I started racing EVs in 1996, and even more when I began driving in EV races in 1998. Even the most jaded, experienced development engineers sit up straighter when you describe getting sideways at the end of a half mile straight because your regenerative braking kicked in harder than expected. Racing is the flea that bites the tail that wags the auto industry dog.

There are strong precedents for using racing to develop and promote EVs. One of the first sanctioned auto races in the U.S. was held in Cranston, Rhode Island on 7 September 1896. The contest's two EVs placed first and second, ahead of five ICVs. The winner, Andrew Riker, went on to become the very first president of what is now SAE. Riker's name may not be familiar, but you've heard of his VP: Henry Ford.

SAE sanctions a series of annual collegiate design contests, in which students design and build small racecars, then drive them in competition. The most prestigious of these events, Formula SAE, fostered an electrified spinoff, called Formula Hybrid, which first ran in 2007. Since 2013, Formula SAE has included its own BEV class. Because of SAE's foresight, quite a few of tomorrow's automotive engineers are experiencing electric propulsion's design challenges and performance opportunities today.

During the 1990s, spurred by the California emissions mandate later chronicled in Chris Paine's 2006 documentary film "Who Killed the Electric Car?", a group called Electric Vehicle Technology Competitions (EVTC) held races for a wide variety of EV classes. EVTC foundered after 1999, but not before giving rise to two spinoffs. The University Consortium for Electric Vehicle Racing Technology (UCEVRT) allowed university teams to continue developing and racing their EVTC Formula Lightning open-wheel racecars until 2004, when it, too, folded. The National Electric Drag Racing Association (NEDRA), started when a disgruntled group of drag racers broke away from EVTC, remains strong and active at the grassroots level. The quickest BEV dragsters run times and top speeds that would have been competitive with unlimited Top Fuel dragsters in the mid-1960s.

Electrified racecars are showing up in increasingly visible, mainstream motorsports arenas. The Pikes Peak International Hill Climb, America's second oldest race, provides an excellent showcase for BEVs. The thin air at the mountain course's high altitude - 2,862 m (9,390 ft) at the start, and 4,300 m (14,115 ft) at the finish - significantly reduces the power of air-breathing engines but has minimal impact on electric motors. For nuanced reasons having nothing to do with energy security or the environment, the rules for the 24 Hours of LeMans strongly favor charge sustaining HEVs. The effect is so strong, in fact, that the Automobile Club de l'Ouest, which organizes LeMans, was forced to amend the rules, handicapping HEVs attempting to keep conventional ICV racers competitive.

We may have needed Tesla to show the public that EVs aren't just golf cars, but it's not for lack of effort on the part of racers. I've had the honor to work with quite a few student racing groups over the years. This all-girl team converted a Lola Formula Ford to run in EVTC's Formula E class in 1998. The car's number, 16, reflects their age. None of the girls had a racing license, so I got to fill in when their scheduled driver was

Figure 6 - Lolita! The Highland Thunder all-girl race team from Port Hawkesbury, Nova Scotia, with the crotchety old man they allowed to drive their converted Lola in Formula E at Phoenix in 1998.

unable to compete. Slower than our better-funded competitors, we won 1st place with clever strategy.

As of this writing (early 2017), there is an active international series for electric formula cars (also called Formula E), with a series for modified Tesla sedans (Electric GT) in the developmental stages. Other racing niches remain unfilled, providing opportunities for creative entrepreneurs. In particular, I see a need for a participant-focused series that would allow hobbyists broad design latitude, both to convert existing cars and to build original designs from scratch, running 100 km road races with no pit stops. SCCA and NASA (the land-based one), are you listening? In any event, here's hoping that all the existing series will succeed, and that an increasingly broad variety of EV racing will arise to excite the passions of the next generation of enthusiasts.

The Path to Zero

As we strive for sustainability, we need to recognize that perfection is the mortal enemy of progress. Analysis is necessary to point us in the right direction, but there comes a time when we just have to start down the path. When you analyze as much data as I have, for as long as I have, it becomes crystal clear that North Americans seeking a more sustainable transportation system should take four simple steps: 1) reduce demand for motorized travel, by replacing motion with forethought; 2) match vehicles more closely to their loads, by either downsizing the vehicles or consolidating the loads; 3) electrify vehicle powertrains to the maximum reasonable extent, as dictated by the vehicle mission; and 4) collect transportation energy from renewable sources, with an emphasis on distributed PV. Long time sustainability advocates will recognize a familiar pattern here: "Conservation enables renewables."

To understand and remember these steps, it's helpful to picture a motor scooter and a PV panel. Though there's probably some nutcase exception who proves the rule, no normal person is likely to make a 100-mile daily commute on a scooter [Step 1 - reduce demand]. A scooter carries one or two people, and maybe a bag or two of groceries [Step 2 - match vehicle to load]. Fully highway-capable electric motorcycles consume under 100 watt-hours of energy per mile traveled (Wh/mi), about a third of the consumption of gasoline scooters; electric scooters consume even less [Step 3 - electrify]. In western North Carolina, a 285-watt PV panel with a microinverter will put over 1100 Wh per day on the grid, annual average. [Step 4 - renewables]. Conservatively, taking these four steps, to this level, would reliably provide over ten miles per day of clean transportation, at a price most Americans can easily afford. OPEC has no say in the matter.

It's important to understand that the scooter and solar panel constitute a thought exercise, not some bitter pill which everybody will be forced to swallow. The U.S. is still a free country, and no one will put guns to our heads if we fail to follow these steps, or don't follow them to this level. If you want to make longer commutes, if you want four wheels and weather protection, if you're not ready to put PV on your roof, or even if you just want to avoid modernity altogether, that's your prerogative. But have a long chat with your conscience before you check out of the game completely, because our energy choices really do have moral implications.

One more thing: Don't sacrifice. Sustainability is like losing weight. If the diet is dismal, or the exercise excruciating, you won't stick with it long enough to make a difference. So, choose only the steps that improve your life, like spending less time in traffic jams, driving a more refined vehicle, and pumping less of your hard-earned money out of your wallet.

Extinguishing the Home Fires

My wife, Beth, and I have been interested in sustainability for decades. Through the years of our marriage, we've managed our yards and gardens organically, and significantly improved the efficiency of the houses we've owned. But we always understood at some level that we weren't going to stay put, which meant that resale considerations often limited the scope of our efforts.

Then, in the summer of 2007, we moved to Asheville, NC. While renting long enough to get the lay of the land, we identified the parts of town where car use could be a luxurious choice, rather than an onerous requirement. In the spring of 2008, we bought a 1400 square foot bungalow, built on a small urban lot in 1955. It has a walk-up attic, a full basement with garage, a south-facing roof, and attractive gardening possibilities. In our previous location in another city, we drove almost 50,000 miles per year; the mere act of thoughtful neighborhood selection in Asheville dropped that to about 16,000. In the ensuing years, the passing of several elderly, out-of-town relatives has cut our annual driving further, to about 8,000 miles. Per capita, Americans average well over 12,000 miles per year. At a third of that level, it's safe to say that Beth and I have made significant progress on Step 1 (reducing demand for motorized travel).

Shortly after buying the house, a powerful new realization dawned on us: we're not leaving here under our own power, at least not willingly. We love our house, our walkable neighborhood, our town, our region. More than anything, we love our neighbors. John Denver's lyric finally made first-hand sense: "we'd come home to a place we'd never lived before".

The rooted sense of permanence is a luxury we'd never known as nomads. We no longer consider what value others might attach to our renovations; if a change makes sense to us, we make it. So, our move toward sustainability in Asheville began like it had everywhere else, with chemical free "yardening" and easy efficiency upgrades, but it quickly went much deeper. We installed a light colored, standing seam, metal roof, turning the attic into a much more useful space by drastically lowering the summer temperature. The standing seams also simplified PV installation when we finally got to that step.

The house came with an odd combination of systems, including both a central, forced air heat pump and baseboard radiator heating. The original oil-fired boiler had been retrofitted to burn natural gas, but with no modifications to improve its efficiency. The kitchen stove and the domestic hot water (DHW) heater also burned gas. In the first few years of living with the house, we figured out that it made sense to use the heat pump during spring and fall, when it was unlikely to employ the resistance heat strips. In really cold weather, the boiler was more efficient, so we used it. We run air conditioning only when the temperature gets really high for extended periods. Fortunately, such times are rare in the mountains of western North Carolina. Opening the windows at night provides adequate cooling for all but the hottest week or two of the summer.

Our goal from the beginning was to eliminate all combustion, not just fossil fuels. In a world populated by over seven billion people, using wood or any other biomass for fuel is not broadly sustainable. Photosynthesis is an essential and effective process for collecting solar energy, but it's extremely inefficient. At typical photosynthetic efficiency levels, land (solar collection area) becomes very dear, and food and fiber outrank fuel as land use priorities. It made sense to us to shift to a totally electric house, then use PV for energy production. Besides being orders of magnitude more efficient than photosynthesis, PV can be placed on roofs, parking lots, and other spaces which are already occupied, requiring no new commitments of land area. LED lights and a more efficient refrigerator were easy, relatively inexpensive, drop-in steps toward the goal of a totally electric house.

Replacing the gas boiler and conventional heat pump with a ground source heat pump (GSHP) reduced our annual household energy consumption by at least 50 percent, from well over 20,000 kWh per year to about 10,000. Because electricity costs more per kWh than natural gas, cost savings weren't as pronounced as energy savings, going from about $1,300 to about $1,000 annually, roughly a 25 percent reduction. [Side note: Weather varies significantly from year to year, which makes precise comparisons challenging. All savings estimates presented here, dollars and energy alike, are intentionally conservative.]

The process of phasing out natural gas was informative, in a geeky way. After installing the GSHP, we kept our gas stove almost two years, before replacing it with electric, and used gas for DHW another year beyond that. The electric stove doesn't appear to have made much difference in our overall (gas plus electric) energy consumption. But, with DHW as the last, lone gas appliance, the fixed $10.00 monthly service charge constituted almost half of our gas bill.

The options for DHW were surprising. Solar thermal collectors, at typical efficiencies of 50 to 65 percent, are 2.8 to 3.6 times as efficient as modern PV panels, at about 18 percent. Not long ago, solar thermal would have been the only logical way to heat DHW in a net zero house. However, recent reductions in PV prices have enabled another possibility. Heat pump DHW heaters achieve typical coefficients of performance around 2.5. At a "quick and dirty" level of approximation, a combined PV plus HPDHW system requires only slightly more roof space (collection area) to produce the same DHW as solar thermal.

Installation is simpler for the combined system, involving only a few extra PV panels, an additional breaker in the service panel, minimal plumbing changes, and no roof penetrations. The most important difference is in system sizing. Achieving net zero electrically requires only that PV production average out to meet demand on an annual basis, whereas solar thermal collectors must be significantly oversized (compared to average demand) to provide sufficient supply in the dead of winter. Thermal overproduction during warm weather or vacations is wasted, and may even require dissipation, while surplus PV production shows up as an energy credit on the electric bill. For these and other, more nuanced reasons, we chose HPDHW, which appears to have cut our DHW energy consumption roughly 60 percent, from 4500 kWh per year to 1800.

Several renovation steps along the way have resulted in a tighter, better insulated house, but we're far from finished. We continue to find opportunities for improving both the efficiency of the building envelope and our lifestyle within it. Sustainability is a work in progress, and always will be.

With annual energy consumption reduced to around 7500 kWh, we set our sights on supply. The original plan was to install sufficient PV to approach net zero with the house alone, then buy an EV and drive it for a couple years, then install additional PV to hit net zero in all but the worst-case years. Using the National Renewable Energy Laboratory's PVWatts calculator (available online at pvwatts.nrel.gov) we predicted that a 5985-watt system (21 panels, 285 watts apiece) would generate 7443 kWh per year while fitting nicely on the available roof space.

Sometimes it's nice to be wrong. In its first year of operation, the system generated about 9400 kWh, over 25 percent higher than predicted. It will be interesting to see how that number varies with year to year weather fluctuations, but we expect to be able to charge our EV with no additional panels.

Because Duke Energy zeroes out any surplus energy balance at the end of each May, we "donated" 2781 kWh, worth $294 at retail, in that first year to a utility which (like most of them) routinely claims that net metered PV owners are freeloaders, sponging off other ratepayers by using the wires

Figure 7 - Looking north from the back yard. Only the top few inches of PV are visible from the street. Photo courtesy of Sundance Power Systems.

without paying for them. Contrary to their disingenuous spin, we pay Duke $11.13 a month for basic service and, ironically, another $1.30 a month to compensate them for their expenses in meeting the Renewable Energy Portfolio Standard. None of the REPS money comes back to us, despite the fact that we had to sign over all of the Renewable Energy Credits we generate in order to be approved for net metering. With $0.87 tax, we pay Duke $13.30 a month, $159.60 per year to use the wires, plus that 2781 kWh "donation" (which, for the sake of intellectual honesty, I'll admit will drop to about 1200kWh with EV charging). Freeloaders, my eye.

Finally, the EV Grin!

I've worked with HEVs since 1986, and PEVs since 1991, but Beth and I aren't early adopters. We don't buy a new car until the old one stops serving our purposes. So, in spite of having worked on fifty or so EV development projects, about a dozen of which I can properly call "mine" (paid for in blood, sweat, tears, and time), we didn't buy our first PEV until July 2016, a year after our PV system was installed. It was a gently-used 2015 Nissan LEAF, eleven months old, with 9,000 miles, at a very affordable price.

Always looking for a chance to promote the PV/EV symbiosis, we got a vanity license tag: SOLRCAR, on North Carolina's "First in Freedom" background (which seems particularly appropriate). Seeing the plate, more than one person has commented on

Figure 8 – Erb's with their Nissan LEAF, Solar powered EVs really are "First in Freedom." Photo courtesy of Sundance Power Systems.

the car's lack of PV panels. This has allowed me to espouse a long-held view: the only Swiss Army knives I want are red and fit in my pocket. Even then, I prefer the smaller and simpler versions. Amphibious aircraft are lousy airplanes and worse boats. Similarly, true solar cars (the spindly photon torpedoes that run in Sunrayce and the World Solar Challenge) are lousy cars and worse energy solutions, no matter how educational and fun they may be for my students and society. PV panels belong on stationary structures, where they can always have optimum solar exposure, and on the grid, where their output is still useful after the EV's battery is fully charged.

The LEAF has been well suited to our needs, as expected. Even though we have the small battery (24 kWh), short range (84-mile EPA rating) version of the car, it easily handles about 6000 miles of our annual driving, roughly 75 percent of our total. This statistic echoes the experience of the first-generation Chevy Volt fleet, which GM's OnStar data show traveling 80 percent of its miles on electricity.

In my work, I frequently compare different types of energy, almost always with the goal of reducing the total amount used. From that perspective, it makes sense to look at energy intensity, where smaller is better, rather than fuel economy, where more is less. Consider a Toyota Prius, with fuel economy of 50 miles per gallon (mpg). Turning that number upside down, the Prius consumes one fiftieth of a gallon per mile (0.02 gal/mi). But not all gallons are created equal, because some fuels are denser than others. All hydrocarbon fuel consumption ultimately reduces to how much hydrogen gets turned into water and how much carbon gets turned into carbon dioxide. Denser fuels contain more hydrogen and carbon (hence, more energy) per gallon. Furthermore, some of our choices (natural gas, electricity) aren't measured in gallons. So, we restate the Prius' fuel intensity to reflect the fact that a gallon of gasoline contains 33,700 Wh of energy, and say that its energy intensity is 674 Wh/mi.

Note that this number represents watt-hours of energy consumed per *vehicle* mile traveled (VMT). VMT is a useful metric, widely employed by the engineers and planners who design infrastructure. However, if we're evaluating options like mass transit, we need to account for the fact that VMT is still slightly removed from the fundamental demand for motorized travel. In that case, we'd make two more adjustments, one for the number of seats in our vehicles, and one for our success in filling those seats (the "load factor"). Then, we could say that a Prius, carrying only its driver, consumes 674 watt-hours of energy per *passenger* mile traveled (Wh/PMT). In a four-person carpool, the same car has an energy intensity of 169 Wh/PMT. But, I digress.

On the EPA combined cycle test, a LEAF consumes 290 Wh/mi, 43 percent of a Prius', equivalent to 116 mpg on gasoline. Because most people think in terms of fuel economy, rather than energy intensity, the dashboard display on a LEAF shows miles per kWh; the EPA test result would display as 3.5 mi/kWh. In the time we've had our car, we've never averaged worse than 250 Wh/mi (4.0 mi/kWh), and warm weather driving has been closer to 200 Wh/mi (5.0 mi/kWh). With our PV system generating over 1500 kWh per installed kW per year, fewer than four of our 21 panels supply 6000 miles of annual driving.

We expect to travel about 2000 miles per year on trips that would be impractical in the LEAF, during which our other car (40 mpg) will consume about 50 gallons of gas. Considering that the average American, including children, consumes over 400 gallons per year, 50 gallons total for two adults doesn't seem too bad. Nonetheless, we look forward to reducing our consumption even further, and to a day when enough people drive PEVs to remove the last trace of loyal soldiers' blood from every tank.

The Case for Optimism

Throughout the four decades that I've been paying close attention, progress towards mass EV adoption has been characterized by a halting herky-jerky; three steps forward, two steps back. The same has been true for renewable energy. But the world's movement toward environmental sanity has been underway for a long time and enjoys broad-based support. Where solar powered EVs are concerned, there's plenty of reason to believe we're finally at a point where any future detours will be minor, as long as we do our part to stay on course.

Jim Clark and Team Lotus, sometimes-ridiculed immigrants from Euro-centric Formula 1, may have been anomalies at Indy in 1963, but they didn't stay that way for long. Lotus, A.J. Watson, and other teams ran mid-engine cars in 1964, with Rodger Ward finishing second in a Watson entry. Then, in 1965, assisted by another group of outsiders (NASCAR's legendary Wood Brothers pit crew), Clark and Lotus finally broke through to win what is arguably the most important race in the world. And, though one lone, quixotic diehard (Jim Hurtubise) kept hoping and striving long past the point of rationality, no front-engine roadster has been even remotely competitive at the Brickyard since.

It would be almost impossible to overstate the enormity of this change. The Indy roadster was the ultimate evolution of a basic automotive layout that predates the Model T. The mechanics who built the roadsters were craftsmen of the highest order, artists in steel, magnesium, and aluminum. Tough, blue collar brawlers like Parnelli Jones had biceps bigger than my thighs, with attitudes to match, on track and off. Green, the official British color in international racing, had long been considered bad luck on U.S. racecars, taboo at American tracks.

The Lotus didn't transform Indy because it was green, or foreign, or even because it was exotically attractive to impressionable seven-year-olds. Neither did Jim Clark, a soft-spoken Scottish gentleman farmer, win because he was polite. The physics of a streamlined, balanced, mid-engine design are simply too compelling to ignore. Engineers like Colin Chapman speak an elegant, calculating poetry all their own, and employ their own talented artisans to give it form. A well-engineered setup changes race driving from a wrestling match into a test of aerobic capacity and thermal endurance, perfectly suited for slender, fit athletes like Jim Clark. And superstition, no matter how widely held, is just ignorance with a backstory.

Any technology as well-matched to its time and task as the Lotus will inevitably prevail. The tipping point, reached at lightning pace in the meritocracy of a racetrack, arrives a lot slower and harder fought in a world where economic incumbency and societal inertia confer the power to hobble competition. But whining about unfairness is a waste of irreplaceable time. Driving on sunshine is every bit as compelling as the mid-engine Indy car, and real companies have already stepped up, happily selling PV and EVs in non-trivial quantities. Which means that today, just like yesterday and tomorrow, advocates for rational transportation have a full day's work ahead of us. If you've seen the future, too, recognize that you're on the right side of history, and remember the words of Andy Warhol: "They always say that time changes things, but you actually have to change them yourself." So, don't just sit there. Roll up your sleeves, and whistle while you work.

Chapter 2 - The Motorcycle Enthusiast   by Mark Bishop and Marsha Livingston

I love motorcycles, tinkering, and finding a good bargain. In 2011, I found a deal on a used 2007 Vectrix VX1 that needed some TLC, and it was a match made in heaven. If the weather is even close to rideable, you will see me about town on it. This was my introduction to electric drive, and I caught the EV bug, bad. I went on to restore and convert a Morris Minor to all electric drive for my brother and I converted a Honda Insight to all electric drive for myself, to use when the weather is less than ideal.

Figure 9 - My restored Vectrix bike with 80-mile range.

Quick Facts:
Net Energy Emissions – 16,600 lbs. - 77% less than similar neighboring homes
Home – Traditional new construction, 3300 sq. ft. in Northeast Tennessee, USA
Technologies Employed – Solar PV, Electric Vehicles, Geothermal Heat Pump
Annual Energy Costs – $300 down 95%

I have an engineering mind and am attracted to things that are simple and efficient. As a teenager, I had a wonderful mentor, Simmie Sanders, who showed me how to craft and repair things with my hands. I was fortunate to marry a woman who shared my passion for building things. We work well as a team and together built a workshop and our home. Prior to undertaking these two projects, I took a course in solar home construction, which greatly influenced what we did.

The Workshop - In 1982, my wife and I purchased 10+ acres in Clinton, TN on a gentle southern-facing slope. We built a 48'x60' workshop and lived in a corner of it for 10 years while we saved for our house. For aesthetics, the shop was built in the shape of a barn with a gambrel roof. We faced one side of the roof exactly south so that we could add solar panels in the future. The living area was 870 sq. ft. or about 30% of the total space. It was well insulated with a central wall built of block and stone for thermal mass. We heated it with a wood-burning stove. Ample trees on our property provided plenty of deadfall for fuel.

Figure 10 – Garage, workshop, and temporary living before the house, solar was added 30 years later, but planned for initially.

Today our former living quarters is the nucleus of my shop, which is used extensively for home projects, several hobbies, and helping friends and neighbors. I used the shop to make all the brackets to mount our solar panels from scrap aluminum I found.

It still has fluorescent lighting, along with many power-hungry tools such as an air compressor, welder, lathe, mill, various saws, drill press, and a planer for woodworking.

Figure 11 - We manufactured our own brackets for solar.

The House - In 1990, we started building our traditional story and a half, 3300 sq. ft. house. We sacrificed the efficiency of passive solar design for a traditional look. However, we incorporated **many** energy saving components. The windows are either double-pane low-E or triple-paned with an R rating of 3.3. The walls are six inches thick with R19 fiberglass insulation, plus a ¾ inch thick foam outer sheet providing a total wall insulation of R25—about double normal insulation in the area. The ceiling has R40 fiberglass insulation, and the floor is insulated to R30. We installed a geothermal heat pump. The geothermal loop consists of four 200-foot-deep wells next to the house. There are two 2-ton heat pumps located in the crawl space under our house—one conditions the upstairs, the other the downstairs.

Figure 12 - Drilling the geothermal heat pump wells.

We incorporated a heat recovery ventilation system that saves 75% of the normal energy lost to air infiltration. This system required us to seal off the inside walls and ceiling with three-ply plastic sheeting, which is sealed at the seams and pictured below.

Figure 13 - Wrapping the interior walls to reduce air infiltration.

The plastic sheeting was applied before we added the sheetrock. All utility boxes on exterior walls were carefully sealed. The entire house is electric, and we are in the process of changing all lighting to more efficient LEDs.

Figure 14 - Mark's restored John Deere in front of the house. The John Deere is one of the few gas users left on the property.

For those planning to build a new, energy-efficient home, here are a few suggestions. In hilly areas, it's best to build on a south-facing slope to take advantage of solar energy. Buy your lot in an area that minimizes daily travel distances. Then, design a home that requires the least energy to sustain it. Minimize exterior surface area with a design that is shaped more like a cube than is the traditional long, narrow rectangle unless the house is designed for passive solar. Incorporating passive solar is the most cost-effective use of solar energy. ***Note***: *We did not follow these passive solar tips for our home.* Plenty of insulation on external surfaces, especially the ceiling, is cheap and pays big dividends. Controlling air infiltration is also very important. A heat recovery system, which we installed, is best. Quality windows and doors with low air leakage rates and high R values are worth the extra cost. Photovoltaic solar systems are expensive up front but pay off in the long run. After the initial payback period, the generated electric power is free. Many solar installation companies have creative financing to spread out--and even eliminate--initial costs.

Why? - In 2008, I had two separate conversations with my brother and my brother-in-law that had a profound effect. They informed me that consuming fossil fuels create air pollution by emitting CO_2 and other compounds. When CO_2 is introduced into the atmosphere, it remains until it is absorbed many years later--primarily by plant life. They added that the ever-increasing levels of CO_2 have an insulating effect, causing the earth's temperature to rise. Then I read a book, <u>Plan B</u> by Lester Brown. It dawned on me that we are severely damaging the quality of life on the only home we have, Planet Earth. I decided to reduce my use of fossil fuels and to show others that they could do the same.

Solar PV - In November 2012, we installed a 10 kW photo voltaic array on the roof of the shop. In preparation for doing so, we installed a new metal roof. It is a 'standing seam' type that allows one to easily clamp the array onto the metal without having to drill holes through the roof. Afterwards, I discovered that there were no 'S5' clamps available that would properly attach to our standing seam, so I designed my own. I machined about 250 clamps from surplus aluminum obtained from the local vocational school. Then my wife and I attached the clamps to the roof along with some aluminum bar to secure the Enphase micro inverters. On top of that, we installed forty-four 240-watt Sharp solar panels. Because we were early adopters, the Tennessee Valley Authority pays us a premium for the green energy these panels produce. Although we only produce 55% of the total power that we use, our power bills are slightly less than $0, thanks to the premium on our green power. How cool is that for the power company to pay you?! Unfortunately, the TVA no longer pays a premium to new solar customers.

Figure 15 - A picture of our electric vehicles and plug-in lawnmower, with our 10 kW solar array on the roof of the workshop.

Transportation – As I began my working career, the world was moving into the electronics age. On my job at an electrical power company, change was coming fast. At first, I was unimpressed with the reliability of the new processor-driven electronic equipment. The industry was determined, however, to make the new technology work. And the reliability improved remarkably. At work and in my shop, I became impressed by the efficiency, power, smoothness, and reliability of 3-phase electric motors and the development of reliable variable-frequency drives to control the motor speed. I then realized the use of 3-phase electric motors in electric vehicles was practical. Coupled with advances in battery technology, electric vehicles are on the verge of becoming mainstream. Here are some advantages of an electric vehicle:

1. A threefold Increase of efficiency, resulting in much lower fuel costs and less environmental damage.
2. Increased safety. A properly designed electric vehicle gives one a greater crumple zone for protection in front and rear end collisions. There is a reduced rollover tendency and greater side impact protection due to low mounted batteries.
3. An improved driving experience, as the vehicle is smoother, quieter, handles better, and has regenerative braking, allowing single pedal driving and saving on brakes.
4. Convenience. Most of the time, there are no refueling stops. It takes just 10 seconds to plug in when I get home, and I wake up to a fully charged vehicle every morning.
5. Very little maintenance. Electric motors have only one moving part. The transaxle is greatly simplified, with no extra gears, clutches, and sensors. An EV has no oil, no drive belts, no exhaust, and no hoses. Other than rotating the tires and checking the brakes, there is little else to do.
6. Significant cost savings. If you pay for your electricity, you can drive three to five times as far as a gas-powered car for the same money. If your electricity is supplied by solar or wind, you essentially drive for free!

My first foray into electric vehicles was a 2007 Vectrix VX1 scooter developed by Lockheed Martin in 1996. The bike is beautifully designed, but in 1996, the scooter used lower power nickel metal hydride batteries, which were failing. I purchased the half-working bike in 2011 hoping to improve it. I ended up upgrading the battery to lithium iron phosphate, transforming the Vectrix into a faster, longer-range machine. It has been totally reliable, and I doubled the range of the initial design from 40 to 80 miles. The only maintenance has been the usual tire changes and tire pressure monitoring.

With the scooter working well, I looked for another conversion project. I was attracted to the 2000 Honda Insight because of its aerodynamics, light weight, and durable construction materials: aluminum and plastic. I replaced the hybrid gas-electric motor with a pure AC electric motor and controller from HPEVS Company. I retained the manual transaxle but removed the flywheel, clutch, pressure plate, and all gears except for second gear. Therefore, the car drives and behaves as one with an automatic transmission or any other EV. The battery is a surplus 2010 Nissan LEAF lithium unit mounted directly behind the seat in the same area as the former gas tank. I added an electric air conditioner and an electric heater. After removing the exhaust system and fuel tank, I added a belly pan, enhancing its already low coefficient of drag. Except for a problem with the mechanical coupler connecting the motor to the transmission—a design and fabrication issue--the car has been remarkably reliable. The car has a large trunk space that has helped me keep my gas-powered pickup truck parked.

Energy consumption is roughly 160-watt hours per mile, which is 30% to 100% better than current mass-produced electrics. The range is about 100 miles, so the Insight is best suited for day-to-day use around town, for which it is a delight. But I did manage to drive it on a 375-mile trip to an Electric Vehicle Conversion Convention in Cape Girardeau MO.

Our third vehicle is a 2014 Tesla Model S85. There is plenty of information on these cars via the internet. We have had practically zero problems with this car and we're extremely happy with it. We think it is the best car that one can buy. This replaced my wife's Cadillac Deville, and we've never looked back. The smoothness, handling, and acceleration of the Tesla Model S keeps my wife smiling. The low running costs are what keeps me happy. We have taken several trips and the expanding Supercharger network makes this easy to do.

Figure 16 - Marsha in her all-electric Tesla Model S.

Annual mileage and energy consumption rates for our vehicles:

> 2007 Vectrix scooter, 6000 @ 100-watt hours per mile
>
> 2000 Honda Insight, 9000 @ 160-watt hours per mile
>
> 2014 Tesla model S, 9000 @ 308-watt hours per mile

For us this translates to 26,000 miles of driving per year at a fuel cost of about $490, **if** we were to pay retail electric rates. Imagine driving 2,300 miles per month for $44. But for us, with our solar credits, the cost to drive is $0. Driving electric vehicles provides a huge energy savings, as they consume roughly one third the energy as a similar gas-powered one. And they are far easier on the environment. They are truly delightful to drive, with maintenance requirements that are not much more than that of a cell phone. By 2018, there should be several affordable models available that boast ranges of 200 miles or greater. At least one model will have a nationwide, fast-charging infrastructure in place as well.

In addition to electric cars, we have an electric (battery powered) lawn mower to trim the yard.

In conclusion - The annual energy consumption per year for our entire compound--home, shop, and our electric vehicles--has averaged 24,296 kWh over the past two years. We estimate the home uses about half the power of any of our neighbors. We estimate the electric vehicles and the shop each consume about 25% of our power. Our annual solar generation has averaged 12,463 kWh for the past three years, or just over half of our usage. We have a 10-year contract with our electric company, Tennessee Valley Authority, that gave us $1000 upon initial startup, and pays $0.12 **plus** the retail rate per kWh that we generate. We also have a contract for the following 10 years where they will pay the retail rate for our generation. We have recouped 55% of our initial investment in the photo voltaic system in the first four years. We have not paid an electric bill since December of 2012, and trips to a gas station are now VERY rare. Thus, we are very pleased with our experiment to cut emissions, simplify our lives and save money.

Chapter 3 - The Green Guy

by David Hrivnak

When two Hrivnaks relocated to NE Tennessee to work at Eastman Chemical, people found a way to tell us apart. My brother has talent and is a good singer and dancer, whereas I have two left feet. So, I became known as that "Green" guy.

Quick Facts:
> **Net Energy Emissions** – 7,200 lbs. - 83% less than the typical American family
> **Home** – New passive solar build, 2,600 square feet in North East Tennessee, USA
> **Technologies Employed** – Passive solar, solar PV, electric vehicles
> **Annual Energy Costs** – $750, down 77%

Figure 17 – The Hrivnak family has reduced emissions from 7% above average to 83% below average.

Being active in the Boy Scouts as both a youth and adult leader, I developed a strong respect for the environment. I want to do what I can to leave my corner of the world better than I found it. As a Christian, I am drawn to Genesis 2:15, where God says, "The Lord God took the man and put him in the Garden of Eden to work it and **take care** of it." To me, this speaks to the need for active care of our planet.

I grew up in the 1960s in Pittsburgh, Pa. At that time, Pittsburgh was an industrial city, and steel was king. Back then, a "clear" day meant brownish grey skies and brown rivers. Between 1868 and 1969, the Cuyahoga River in nearby Cleveland was so polluted that it caught fire thirteen times.[14] Lake Erie was devoid of fish[15] and so polluted that my parents would not allow us to swim in it. We were relegated to making sand castles on the beach.

The clean air and water acts of the 1970s did a remarkable job of repairing the environmental damage to the area. We now swim in Lake Erie, fish in the Cuyahoga River, and Pittsburgh has frequent clear-blue skies. Rather than sparking financial ruin, the economy is now stronger. Yes, there was change and disruption and not everyone came out ahead, but most did. Pittsburgh is now noted for software development and medical research rather than steel. The city survived, and, in 2010, Pittsburgh topped Forbes' list as the most livable city in America.[16] Both Pittsburgh and the United States have shown that the careful application of environmental policies benefits us all.

The final event that motivated my concern for the environment happened in 1974. Just as I was getting my driver's license, the first energy crisis hit. I remember spending hours in gas lines and switching license plates so I could buy gas on the right day as gas prices tripled. Since then, I have done my best to be easy on the environment and become self-reliant. As a result, I have built two passive solar homes and developed several energy-related hobbies.

We now produce our electricity through our rooftop solar system, which powers the house and both plug-in cars. Being Net 0 has been a goal of ours, and we worked on a phase every year or two. Through this, we have proven that a significant reduction in emissions is achievable, while still maintaining a comfortable lifestyle. The three major areas we focused on were transportation, rooftop solar (Photovoltaic) PV, and conservation, mostly LED lighting, in addition to a passive-solar home design. This reduced our CO_2 emissions over 19 tons per year from our starting level and our carbon footprint by 85%.

Transportation – As with many families, transportation was our largest emission source. With people changing cars more often than homes, it is also the category where we can make the fastest change. We used to burn 1,200 gallons of gasoline a year. The biggest offender was our Chevy Avalanche, which was used for family trips and frequent camping trips with the scouts. The Avalanche averaged 16 miles per gallon (MPG), while our second car (a Geo Prizm) got a respectable 28.5 MPG.

When gasoline prices went above $4 a gallon, I began to experiment with electric drive by converting the Avalanche into a plug-in hybrid. The electric motor helped push the truck whenever I applied the accelerator. I learned much about the potential of electric drive and lithium batteries, but, in hindsight, it was a large investment for a relatively modest gain in fuel economy. However, through my experimentation, I caught the electric vehicle (EV) bug and went on to convert a Jeep to all-electric drive. With the new lithium ion batteries, I realized EVs were becoming practical, and I wanted to be part of this new wave of cleaner transportation. When I finished converting the Jeep and had to give it back to its owner, I started to research what EV to build (or buy), as I was having EV withdrawals. Once you get hooked on the smooth, quiet, instant torque, it is hard to go back to a gas car. In my search, I

found several used Tesla Roadsters for sale. While I knew I could not afford a new one, the prices had come down on used models to make ownership possible for me. I convinced myself that if I were going to convert more vehicles, I should learn from the best EV available. The Tesla Roadster is expensive, but it outruns and out handles the more expensive Porsche Turbo Carrera and Ferrari 550, and it sure turns heads.

Figure 18 - Tesla Roadster shown at a school energy talk.

The Tesla Roadster opens eyes to the possibilities of fun, clean, electric drive. Here is a sharp-looking car that travels over 300 miles on a charge. Fuel costs are about half that of a Prius, yet you can take the Roadster to the drag strip and humble most any performance street car. One evening at Bristol's

Thunder Valley drag strip, in a drag racing event known as "Street Fights," I lined up against three different Corvettes, a Shelby Mustang, and BMW M5 sport, only to watch them in the rearview mirror. With instant torque, the Roadster jumped off the line, and no one could catch me in the quarter mile run once the green light came on. Here is super car performance, with emissions less than the best-in-class gasoline Prius, and **much** less than a similarly-performing sports car. I have taken the Roadster to many car shows, environmental conferences, Earth Day events, and local high schools, and it always draws a crowd. I see the genius of Elon Musk's plan to build compelling cars that people want, not because they are "green" but because they are the best cars available.

We then added a plug-in Volt as my wife's daily drive and for long trips. This significantly dropped our gasoline from 1,200 gallons to 225 gallons per year. Thanks to the EV tax credit, the new Volt was less than a similarly equipped Prius. Nearly half of the Volt's miles is in electric drive, powered from solar. On out-of-town trips, the Volt's gas engine kicks in at 40 mpg, accounting for our gasoline usage. This reduced our CO_2 emissions from 7% above average to 45% below average.

The drop in fuel usage also saves a noticeable amount of money. At $2.50 a gallon, the fuel savings is $200 per month. To be fair, we use about 260 kWh a month for our gas-free driving. That electricity would cost us $25 a month if we had to buy it, offsetting some of the savings. But we no longer see those costs, since we make our own "fuel" with rooftop solar. Most charging is done at home--another advantage of EV driving--as we start each day with a full battery.

One complaint I frequently hear is that EVs are expensive. Thanks to the federal electric vehicle tax credit of $7,500, EVs are affordable. A $37,000 Chevy Volt drops to $29,500. The Nissan LEAF starts at $29,000 and drops to $21,500 with the tax credit. While I am not a fan of tax credits, they are significantly less costly than the Gulf Wars we fought over oil.

Solar PV – As we electrified our vehicles, we wanted to take sustainability to the next level and generate our own power. While charging an EV from the grid is cleaner than burning gasoline[17], it is not emission free, but if you power the car via solar, it is. We started by installing a 7 kW system, using 28 panels. This was later expanded to 8.2 kW, using 33 panels, after purchasing the plug-in Volt. Our local power company supports net metering, allowing us to bring our electric usage down to, but not below, zero over the course of a year.

We still pay a modest connection fee each month, which is reasonable, as we have access to the power grid, which acts as a huge battery, storing any excess power we generate. We generate significantly more power than we use in the

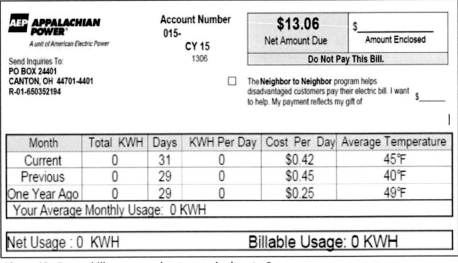

Figure 19 - Power bill, our annual net usage is close to 0.

spring and fall, when we do not need heating or air conditioning. In the summer, we are close to break even, adding up to over 1,500 kWh of excess production from March to October. This is helpful as those "banked" kWh covers our winter shortfall.   One thing I just noticed in the below graph is that summer cooling is now larger than winter heating.   The reverse of when we first moved to TN.

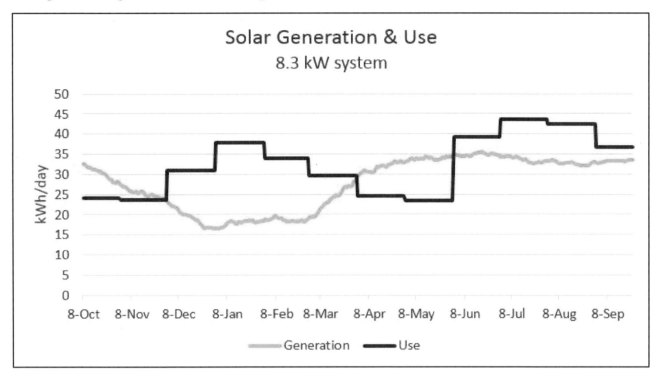

Figure 20 - Solar generation through the year. Our output drops noticeably in the winter.

In Northeast Tennessee, we average about four hours of peak sunlight a day. Thus a 255-watt panel generates 1 kWh per day on average; therefore 33 solar panels generate 32 kWh per day (or about 11,000 kWh per year) of power. As you can see from the graph, there is significant variation through the year. Note: After we purchased the Volt, we added five more panels to maintain our goal of net-zero power use.

Another advantage of our solar set-up is that we have some backup power in the event of a grid failure. The Sunny Boy SMA line of inverters has a secure-power feature, so we can produce power when the sun is shining even if the grid is down. This power is limited to two 1,500W circuits, which are enough to keep the refrigerator running and a car charged during the day.

Figure 21 - Inverters with secure power feature.

During our first year, it snowed, and were pleased to see that the solar panels were the last thing the snow stuck to, and the first place it melted off. In year two, we had another dip with back-to-back snows and temperatures in the single digits. This dropped production to near zero for several days. That

Figure 22 - Snow is a minor issue, as it melts off fast in our area.

was a time I was pleased we remained grid connected, as our usage also spiked due to the extreme cold.

The cost of the solar system was about $20,000 with the federal tax credit. If I get the expected 30-year life out of the system, our power costs should be $.067 per kWh, or 30 percent below current power rates. While not a great "investment," it is better than CDs or bonds and is more stable than stocks.

One solid piece of advice we received as we planned our solar installation was that the best way to save on solar is to first reduce electric usage to a minimum. We found that conserving electric usage is more cost-effective than adding additional solar. For example, spending $300 to switch to all LED lighting costs far less than adding two additional solar panels to power inefficient lighting. Therefore, we swapped all lighting to LEDs, made sure we sealed any leaks around doors, and turned off or added timers to cell phone chargers and other electronics.

The system is working as planned, as our average electric bill, including powering both cars, was under $15 a month last year. The total kWh used last year was less than the monthly average for homes in our area. Additionally, the solar PV system offsets about six tons of CO_2 each year.

Design savings – We estimate the basic passive-solar design aspects of our house save about six tons of CO_2 a year and about half the heating and cooling costs over the average area home. We realize few people can build a new home, but if you do, employing the best green-building design and techniques produce energy savings that go on forever. Careful consideration of orientation, glass placement, overhangs, air infiltration, and insulation pays great dividends over the home's life. Figures 24 -28 attempt to show this in a very visual way.

Basic home design – As an engineer and an avid volunteer with Habitat for Humanity, I have a good background and experience in building. So, when we decided to build, we pushed the boundaries on energy efficiency. We began with a passive solar design, orienting the home and windows to take advantage of the southern sun. As such, most of our glass is on the southern side, with overhangs designed to shade the windows from sunrise to sunset with the high summer sun. In the winter when the sun is lower, we enjoy full sun all day. The house is set on an insulated slab, so the slab and tile warm slowly during the day and release heat through the evening and night, moderating the temperature swings.

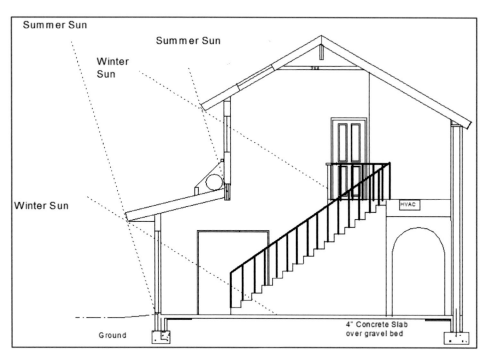

Figure 23 - Diagram of how the sun shines into the home during the winter but not summer.

To show the dramatic changes in sun angles throughout the year, we captured a series of pictures.

In December and January, the sun penetrates deep into the home. This warms the home, even on cool winter days. The sun penetrates deepest on the winter solstice on Dec 21.

Figure 24 - The sun penetrates deeply in December.

In November and February, the sun still penetrates deep, continuing to add warmth through the day, and with the insulation, the heat lasts much of the night. The open design allows the heat to easily circulate from upstairs to downstairs, and we use a paddle fan to keep the temperatures from stratifying, as the central heating system does not run much.

By mid-September and mid-March, the sun is coming in only a third as far as during the winter solstice. This is the sun's position at noon in mid-September and March, the spring and fall equinox.

Figure 25 - Sun position in November & February.

Figure 26 - Sun position in Sept. and March.

In late April and mid-August, only a small amount of sun comes into the house.

Figure 27- Sun barely coming in April & August.

From late May through early August, no sun shines into the home. This picture was taken mid-July around noon. The shading from the overhangs helps to keep the home from overheating in the hot summer months. With no sun hitting the windows, we save on summer air conditioning.

Figure 28- No sun hits windows in June & July.

We have a few deciduous trees on the south and west sides of the house to further shade the home in the fall, when temperatures still tend to be warm, but the sun starts to come back into the house.

If we had most windows on the east and west sides of the house, the opposite would happen because we would catch the summer morning and afternoon sun, significantly adding to summer air conditioning loads, while getting virtually no gain in the winter when we need it most. The solar orientation provides about half our heating.

Along with the passive solar design, we tried several innovative ways to save on long-term utility bills.

We built the north and west walls with Insulating Concrete Forms (ICFs).

The ICF forms interlock in large 16" x 4' blocks. They are made locally out of Styrofoam. The forms snapped together, and we added both horizontal and vertical rebar as we built. We cut openings for doors, windows, and any other penetrations and framed them in with pressure-treated 2-by-12s.

Figure 29 - Insulating concrete forms insulate very well and keeps out drafts.

We filled the hollow core with six inches of solid concrete from a pumper truck. The walls insulate as well as an R50, wood-frame wall of fiberglass insulation. ICFs also have very low infiltration, which keeps drafts out. With the six inches of reinforced concrete, the house is very solid and designed to withstand hurricane-force winds. Additionally, this makes for a very quiet home.

Figure 30 - Filled with reinforced concrete, the ICF's are exceedingly strong,

Figure 31 - Structural insulating panels for the roof.

For a roofing system, we used Structural Insulating Panels (SIPs). Imagine a 4-foot-by 15-foot sheet of plywood, laminated to eight inches of Styrofoam, laminated to another 4-foot by 15-foot sheet of plywood. The SIPs were made locally by the same company that made the ICFs.

The eight inches of foam insulates as well as R 38 fiberglass and, like the ICFs, the SIP's have very low infiltration.

The southern wall was stick-built, with 2-by-6 studs covered with foam insulation. While not as good as an ICF wall, it is better than the typical wall. The windows are low-E and argon filled, thus making the house extremely well insulated by trapping the sunlight and heat that make it into the home. We estimate that the higher levels of insulation save us 25 percent over the average area home. When combined with passive solar, our heating and air conditioning needs are about 70 percent less than the average area home.

While more expensive than traditional construction (an extra $8,700 or $3.30/sq. ft.), because we were our own contractors and did some of the work ourselves, we built the home for $65 a square foot in 2000. At the time, this was on the low end of the $60 -$100 a square foot that contractors were quoting to build similarly sized homes.

As we built, we integrated many other energy-saving tricks such as adding breadbox solar water heating. The black tanks in an insulating box help warm water before going into the hot water heater. On sunny spring and fall days, the water going into the water heater is warmer than the water heater setting, giving us free hot water. (Free, if you discount the cost of the extra water tanks and my labor).

Figure 32 - Breadbox solar water heating.

Recreation – In this part of the country, many people pull trailers for camping, boating, and/or hunting trips. People ask if they need to give up their hobby if they go "green," and the short answer is no, within bounds. However, aside from the Tesla Model X, which is rated to tow 5,000 lbs., most plug-ins are relegated to lighter loads. However, thanks to an aftermarket hitch and lots of low end torque, one can pull a trailer. There are several lighter alternatives, like our ALiner pop-up camper, complete with heat/AC and running water. The dry weight is 1600 lb. and about 2,000 pounds loaded. We added

Figure 33 - Aliner popup is pulled well by our Volt.

a lightweight solar panel to keep the 12v battery charged for longer trips, as many National Parks do not have any connections. Pulling this trailer--or a light fishing boat or canoe, kayak or bicycle--is well within the capability of an EV. As with gasoline cars, the range does take a hit, so one needs to plan accordingly.

Even my Roadster has a hitch I often attach a bike rack to. (Or, as I say at car shows, "I use the hitch to tow gasoline cars that have run out of gas, to the gas station".)

One more Step - As much as we try to reduce it, we still leave a carbon footprint. To help offset that, I am involved with our church, where I created an energy task force to look at ways to reduce energy costs. We started with a free energy audit by the

Figure 34 - We added a solar panel to keep the battery charged in National Parks.

Green Interfaith Community. We realized we had a lot of very inefficient lighting and some old and bad equipment, like a donated freezer that was using $350 of electricity/year. We replaced it with an energy-saver model with a payback of under 2½ years. We have gone on to replace virtually all lighting with LEDs, saving us $4100 and 28 tons of CO_2 a year. Even more exciting is the fact that now many church members are sharing ways to save energy and money at home, thereby multiplying the effect.

I also present to local schools, civic groups, and participate on the National Solar Tour of Homes, National Drive Electric Week and frequent car shows to help get the word out that sustainability is here and now.

Conclusion - How much do you pay for electricity and three-plus tanks of "fuel" for your vehicles each month? Imagine the savings—to your pocketbook and to the environment—for taking the steps we have taken. We have been able to dramatically reduce our costs and emissions by using a combination of energy efficient building, passive solar design, electric vehicles, rooftop solar, and LED lighting. For us, the clean energy future is here now.

As you see, there is no one silver bullet when it comes to energy savings, as **many** items and attention to details add up to significant and lasting savings. On the home front, we estimate the following savings over the typical area home.

HVAC – 70 percent reduction

- Passive solar orientation
- Very well insulated (R-40+)
- High efficiency heat pump (18.5 SEER)
- Low-E Argon windows
- Careful attention to air leaks
- Landscaped for appropriate shade/sun
- Open windows on cool spring/fall nights
- Close insulating shades on winter nights

Lights – 75 percent reduction

- Changed to all LED lighting

Appliances – 30 percent reduction

- Energy Star appliances
- High efficiency washing machine

Miscellaneous – 30 percent reduction

- Turn off lights when not in use
- Timers on chargers
- Insulation blanket on water heater
- Low-flow shower heads
- Battery-powered lawnmower and lawn tools

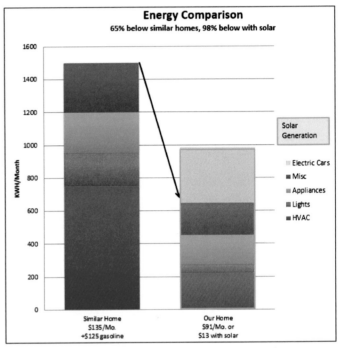

Figure 19 - Our energy use compared with the average home.

Month	Total KWH	Days	KWH Per Day	Cost Per Day	Average Temperature
Current	0	31	0	$0.42	45°F
Previous	0	29	0	$0.45	40°F
One Year Ago	0	29	0	$0.25	49°F
Your Average Monthly Usage: 0 KWH					
Net Usage : 0 KWH			Billable Usage: 0 KWH		

Figure 20 – Our annual power use is now down to 0.

With a 60 percent reduction in household energy usage from the average home, we are easily able to accommodate the power needed for our plug-in cars and remain below average. Using solar brings us close to 0 usage. We use about 13 therms of natural gas a month, which we plan to phase out.

While we paid extra for the insulation, plug-in cars, and solar panels, we look at these as long-term investments in the future. Because of this, we will reap the benefits of near zero energy costs for the rest of our lives. We are energy independent, and our CO_2 emissions are 19 tons per year lower than average. We hope that through this endeavor, we will leave our corner of the world a little better than we found it. Rather than being a sacrifice, our home is quieter and brighter than most, and our cars accelerate and handle better than their gas equivalents. We look at this as a win-win-win-win and look forward to an exciting, clean, cost-effective, and sustainable future.

Chapter 4 - One House, Two Cars, A Quest for Sunshine Symbiosis   by Peder Norby

There's something traditional, in an American sense, about a home and two cars in the garage. We are a nation that came of age the past 100 years concurrent with the era of the automobile. For better or worse our homes and cars are together entwined with the embodied energy of our built history. For better, we now can power both our homes and our cars with harvested sunshine.

Quick Facts:
Net Energy Emissions – (-10,000 lbs.) - 120% less than the typical American family.
We went negative by installing a solar system at a local nonprofit.
Home – New 4,600 square feet in Southern California, USA
Technologies Employed – Solar PV, electric vehicles, passive solar, super insulated
Annual Energy Costs – $245, down 97%

Figure 35 - The beginning, 6 years ago with the Mini-E and our Sungas Station.

We wondered if it is possible to become self-sufficient and really provide all the energy we need. The idea is a simple one: harvest endless sunshine from a small portion of a roof to provide 100% of the energy needed to power a home and two cars with zero utility cost and zero gasoline cost.

To test this theory, we embarked on a one-year trial to document our effort to power our home and two cars, each driven an average of 12,000 miles.

We hoped to make more kWhs than we use over the entire year and become a true zero-emission transportation solution, net zero in use and below net zero in the total cost of energy. We documented all with our utility bills and car readouts and now share our somewhat private information with you. We've done the math, and we've lived this EV + PV life for six years. We've been below zero with our home and one car and believe we can now do it with two cars.

Just as the cell phone, the digital camera memory chip and the computer have transformed how we communicate, advances in technology have led us to an emerging "symbiosis" of the building, the automobile, and the energy plant all working together as a self-contained system owned by a single entity, rather than separate entities at separate locations such as a home, a gas station, and a power-plant. It is a time primed for great change in how we make and distribute energy and how we motor from place to place.

Our personal goal for this project was to save money, to be more self-reliant, to reduce our dependency on foreign oil and its related cost in dollars and lives, and to improve the air quality in our city. Our goal was also to demonstrate this rapidly emerging and symbiotic new energy and transportation future.

Our Home – We live in Carlsbad California in a temperate climate. We were owner-builders of our home in 2006. Our home was awarded the 2008 California Center for Sustainability Energy "Excellence Award" for being a net zero energy home. This award is peer reviewed and goes to one homeowner per year in Southern California. The main home is 3250 sq. ft. plus a 1200 sq. ft. guest home occupied by one. Our home and guest home use approximately 5000 kWh a year of energy, less than half of the average home electricity use in the U.S. The lumber for extra-thick walls, a tankless hot water heater,

Figure 36 - The Norby Home in Southern California.

extra insulation and compact fluorescent light bulbs collectively came to $15,000. We plan to fully recoup this cost, and then some, over the life of the house. Thick walls fashioned out of Douglas fir help maintain the desired temperature. Roof overhangs let sun in during winter, when it is beneficial and screen out the sun's rays in summer. Rolling glass doors allow ocean breezes to penetrate deeply. The "shotgun" configuration, with a continuous hallway from one end to the other, allows those breezes to blow all the way through.

In early 2007 we purchased a SunPower 7.5 kW system that generates approximately 11,500 kWh a year. This system was architecturally integrated into our home at the planning stage and was sized to power the home and one car. This system was completely paid off in utility and gasoline savings in April of 2012. The initial cost was $23,000, which dropped to $11,000 out of pocket with the federal and state incentives.

Figure 37 - Solar, mounted low so it is hidden from view.

In 2014, we added an additional 1 kW of panels for a total system size of 8.5kW, generating 13,000 kWh a year. We did this to accommodate our second plug-in car. Our system is grid connected. We charge our cars at night from the grid when it is less expensive and less taxing to the grid, and we generate extra kWhs during peak hours, providing this energy to our neighbors during peak demand.

We also conserve water. Most of the front and backyard are planted with native shrubs, grasses and trees that aren't very thirsty. The property has 40 different types of manzanitas, as well as Cleveland sage, deer grass and lavender-blooming ceanothus. We have rose and herb gardens, but they take up little space. And our vineyard--we love to make our own wine--is irrigated by pipes that funnel rainwater directly from the roof to the vines. Our water bill is a very low $15 to $20 a month.

Our Cars – Julie and I both drive the fully electric BMW i3. The BMW i3 is one of the most efficient cars and just might be the most efficient car in the world. It is a dream to drive, with leading-edge technology, comfort, and safety.

Figure 38 - BMW I3, Julie's has a range extender.

Julie and I have been field trial drivers of both the BMW Mini-E and BMW Active E for the past five years, and both cars have fit perfectly into our lifestyles, requiring no concessions on our part. I drive approximately 9000 miles a year and use about 2000 kWh per year. Julie drives approximately 15,000 miles a year and uses about 3600 kWh per year.

Of special note, the BMW i3, at 2650 lbs., is lighter than most cars. This lightness will save us over 1000 kWhs of energy each year for the same miles travelled in our older BMW Active E.

The BMW i3 has a carbon fiber body and reminds me of the slot cars I used to play with. I am, once again, just a teenager driving on a real-life slot car track, except that now, the electric trigger/resistor has moved from my forefinger and thumb to my right foot. Go fast, slow down, repeat.

Just like that slot car track of the 70's, my BMW i3 is also plugged into the wall, now sipping the required kilowatt hours of electricity from the sunshine harvested on my roof. Yes, you can make your own fuel; yes, you can drive powered by sunshine.

Unlike any other production car in the world (except for the BMW i8) BMW has pioneered the use of lightweight carbon fiber, CFRP, to make the body of the relatively affordable BMW i3. In my opinion, lightness via CFRP--not battery chemistry or the electric motor--is the secret sauce that has put BMW a full decade ahead of any other car manufacturer in the world. Does it drive like a BMW, even though it's electric and made from CFRP? Heck yes! Let me go farther and say it raises the bar and redefines what a BMW should drive like. The i3 is simply a better and more enjoyable car to drive than its gas brethren.

The are many reasons to favor the electric car: saving the environment, reducing carbon emissions, lowering fuel cost, and eliminating stops at gas stations. The overall driving experience is one of simplicity, quietness, and ease. Once you have tasted the fruit of electric driving in a machine designed for performance, there is no going back to the antiquity of the gasoline engine. Forward to a faster and cleaner future, thanks to the electric motor.

The BMW i3, according to senior BMW executives, is faster than gasoline BMWs to 30mph. It also has one of the shortest stopping distances of any BMW. Quick to accelerate, with a ton of torque, an equally important short stopping distance, and being lightweight are the hallmarks of any performance car, including the BMW i3, which is designed for urban and suburban life. That's not to say the BMW i3 is hard or jerky to drive, as the overall driving experience of the car is one of simplicity, quietness, and ease, surrounded by luxury and quality materials; however, it is a BMW to its core, and loves to be driven in an enthusiastic manner.

On a race track, the i3 is fun, but it's not designed as a track car...correction...it is depending on how you define "track." I recently had the opportunity to drive my i3 on an autocross circuit and also drive many of the other higher tier electric cars on the same course. The car with the fastest time? The BMW i3.

The i3 excels in tight spaces and in the twisty's, as the low weight, fast 0-30 speeds, and short braking distance show its supremacy. I have no doubt that on a longer track with longer straightaways, a car like the Tesla Model S would prevail. The Tesla is a beautiful, sensational electric car, but the relevant question is:

Where is your "track?"

My "track" is the 28-million populated urban jungle of Southern California, where I live; I don't need to drive 250 miles at any one time. Southern California as just one example of EV nirvana, is where the BMW i3 thrives. 28 million people living in a geo area sized roughly a two-hour drive north and south by a two-hour drive east and west--an area filled with thousands of chargers and dozens of fast DC chargers allowing an i3 driver the ability to go and park anywhere they choose.

More importantly, So-Cal is an area that has polluted air due to our love affair with the gasoline automobile. Electric cars like the BMW i3 offer the promise of a better path towards an emission-free future. Think about that with every breath you inhale and our cumulative contribution to the health of that air.

The i3's range is 70 miles on the freeway at 80 miles per hour and 115 miles in the city or a more typical freeway stop and go speed of 35mph. This range has proven to be more than adequate for our lifestyle.

But no two experiences or drivers are the same. For those with longer trip requirements, BMW makes the i3 with a gasoline Range Extender option (REX) thus effectively doubling the car's range and offers the convenience of filling up at any gas station. BMW i3 is unique as compared to any other car in the world, as you can have your choice, a BMW i3 fully electric, or one with a gas REX.

I can't predict the future, but I can drive futuristic cars, and there is no finer representation of what that future looks like than the BMW i3 and i8. And this fun experience saves us money.

Figure 39 - With no motor up front, you can use the space for something more fun and useful.

Our Net Zero Year Experiment – The following are excerpts from the blog we kept documenting our Driving to Net 0 Energy Challenge: http://electric-bmw.blogspot.com/2014/04/one-house-two-cars-quest-for-sunshine.html We documented and shared both our successes and failures. We believe that in just a short decade or so this "Sunshine Symbiosis" will soon become the norm. Solar is getting cheaper, houses are becoming more efficient, and electric cars are getting better, more efficient, and less expensive.

Figure 40 - We created a logo for our experiment.

We began our 12-month Driving to Net Zero journey on May 15th, 2014. We got off to a great start. In the first month we reduced our electric use by 24 kWh and saved $149.20 in utility costs.

Figure 41 - An EV uses 1/5th the energy of a gasoline car.

The first item to tackle in "Driving to Net Zero" is efficiency. It is far cheaper to save kilowatts than it is to make them. Energy efficiency in the home is a very well-known prerequisite for savings. Just like the LED light bulb that uses 1/7th the energy of an incandescent light bulb, we now have cars that use 1/5th the energy of a typical new car to travel the same distance.

In "Driving to Net Zero" an efficient home and the most efficient cars are necessary ingredients. The "most efficient" can also mean the most enjoyable, the most livable, and the most premium to drive, as efficiency in our lives is additive, not punitive.

The second item for " Driving to Net Zero" is computational power, communications, and smart data. If we can measure it, we can improve it.

The terms smart grid and smart house apply here. In the past few years, our utilities have swapped out our "dumb meters" and installed "smart meters" as a move to the future, a future where our smart appliances will communicate under our direction with the energy grid and optimize the times when they operate, thus saving us money.

Figure 42 - Dual ChargePoint station allows us to "fill-up" conveniently at home.

Why is this energy information important? Can you imagine not being able to see your bank account or credit card data? How would you know how to budget and spend/save your money? Energy is no different. When we can see our energy data, we can manage our energy, we can make informed decisions, and we can save money.

An electric car driven 10,000 miles in a year can use 50% of the electricity of a typical house. A household like ours, with two electric cars, can use more electricity for the cars than for the house. With that much electricity use for the cars (still about 1/5 the energy cost of gasoline), it's important to see your overall usage and time of use with a data-connected smart charging station, especially in California which has peak and off-peak pricing.

So, we're off! During this past month, after announcing our "Driving to Net Zero" challenge, we have heard from dozens of homeowners and EV drivers who are on the same path, eliminating their utility and gasoline bills and discovering the beautiful combination of affordable solar PV and electric cars.

Our goal is to promote this way of living and driving as a "better recipe" for our planet as compared to burning fossil fuels in our power-plants and in the less efficient gasoline engines of cars.

……

As we finish the fourth month of the "Driving to Net Zero" energy challenge, I'm reflecting on the beginning of our transition from two gas cars to two electric cars powered by solar. The Solar PV on our house was the first step when we built our home in 2006. Then came the 2008 Mini-E and the fascination about the potential to provide the electricity required by the electric Mini-E from our roof.

Will it work? Will it go up hills? Will the batteries be unreliable? Will it be like a golf cart and begin to slow down on the final four holes? Will it be a toy, or will it be reliable transportation? Will it be boring? Will the headlights dim like a flashlight running out of juice? Can the sun make enough energy via Solar PV to power the car? Will my garage blow a fuse--or worse? Will I be considered a dork in a "play" car?

We kept our two gas cars, as we were unsure the electric Mini-E would work for us. We simply didn't trust it; however, it was an experiment that we were eager to try. We loved our time with the Mini-E # 187 like no other car before. The Mini-E worked beautifully as fun and dependable transportation, far beyond our expectations. After three or four months of not using the other gas car at all, we sold one gas car and became a one electric-one gas car family.

Two and a half years and 35,000 sunshine-powered miles went by fast in the Mini-E. As I was preparing to transition into "my" second pre-production car, the BMW ActiveE, an unexpected and surprising event occurred. My wife Julie claimed dibs on the ActiveE. I decided very quickly it was better for me to drive a gas car and be happily married and let her drive the ActiveE than insist upon driving the ActiveE. I was miserable for several months wanting to get back into an electric car.

After Julie drove the ActiveE for a year without needing a gas car, we then knew that neither one of us needed a gas car and we transitioned to two electric cars. First came the BMW ActiveE followed by the Honda Fit EV enroute to our current garage filled with two BMW i3's. That's the short version of a very long transition from interested but skeptical, to amazed and excited about transitioning to EV's. From where we are now, having completed that transition, it's almost humorous to think about our concerns from several years ago--whether one electric car would work even if we kept the two gas cars. Yet that was exactly our concern and is a concern shared by most who are thinking about beginning their own personal transition from gas to electric. Like us, for most it will be an evolutionary change with many steps in between. The inertia of the status quo (gas) is strong and difficult to change.

So how are we doing in the Driving to Net Zero Energy Challenge, living in a solar powered house and driving two cars a total of 24,000 miles for a year? We're a third of the way through the year with a little over 9000 miles on the i3's. We're still generating more energy than we use, although that will change soon as we get into the winter months. We hope to catch back up and get to zero net usage in the sunnier months of March, April and May.

Last month we saw a large uptick in the energy we used, primarily from driving our French exchange student everywhere and the cooling of our underground wine cellar during a very hot month. The good news was our solar PV generation was also high.

How can we use 117 kWh for the month and have a -$132 credit, you may ask? Simply because energy is cheap at night and expensive during peak hours. Energy is priced $0.49 per kWh during the hours when we produce extra and $0.16 to $0.20 per kWh during the times when we use electricity from the grid. It is very easy to set the timers on cars to take advantage of the off-peak rates.

Norby Home & Two BMW i3's Energy Graph

MONTH	JULIE Miles/kWh	PEDER Miles/kWh	SOLAR kWh	SDG&E kWh	SDG&E Cost	Gasoline Cost
May/June	1046/228	1335/348	1432	-24	-$149.20	0
June/July	1011/231	655/150	1353	-118	-$162.57	0
July/August	1385/261	1035/217	1326	-25	-$156.46	0
August/Sept.	1519/331	1342/274	1435	117	-$131.89	0
Sept./Oct.						
Oct./Nov.						
Nov./Dec.						
Dec./Jan.						
Jan./Feb.						
Feb./March						
March/April						
April/May						
Cumulative	4961/1051	4367/989	5546	-50	-$618.12	0

Figure 43 - Energy use after 4 months into the experiment.

BEFORE SOLAR PV & BMW i

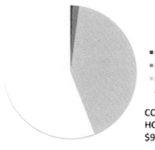

Renewable Energy
Nat Gas Utility
Electric Utility
Gasoline

COST OF ENERGY HOME & CARS $9200 per year

Our prior cars were a Volvo S60R and an Infinity G35. We spent $5000 a year on gasoline and $4000 a year on electricity.

AFTER SOLAR PV & BMW i

Renewable Energy
Nat Gas Utility
Electric Utility
Gasoline

COST OF ENERGY HOME & CARS $245 per year*

We now drive the most efficient cars in the world powered by sunshine. The asterisk on the cost represents the cost of natural gas, which is offset by a larger credit for electricity. Unfortunately the credit is not transferrable with our utility SDG&E.

Figure 44 - Cost savings we have seen by going green.

In month five, we reflected on the economics of our journey, and the result was dramatic for us here in California. Our energy costs have dropped from $9200 to $245, and we believe $0 is within reach.

One parting thought: The air that we breathe is not our private property. It is part of the "great commons" shared by all that inhabit our planet. We should and must be concerned about the health of our commons and our role as private citizens in making them better or making them worse.

Summer is now past as we continue our quest of "Driving to Net Zero Energy Challenge." After five months of living and driving, we have amassed a large electricity credit of $-718 against a natural gas bill of $92, and we have used close to zero (11 kWh) or about $1.50 worth of net electricity.

It's clear to us knowing our past few years of usage, that we will achieve our goal of harvesting sunshine from our roof and providing 100% of the energy needed to power our home and two cars with zero utility cost and zero gasoline cost for a year.

We anticipate accomplishing this goal four months earlier than planned at our annual utility bill true-up in January. That true-up bill will have a big fat goose egg at the bottom with an unused credit of around $450.

The harder goal to reach will be actual net zero energy consumption. We point out there is no financial benefit, only cost, once you have a zero-energy bill, with our provider, any effort to conserve beyond that is a gift to the utility.

We anticipate being within ~700 kWh of this goal at the end of 12 months, so it's going to be very close. If we have a warm and sunny winter/spring, we might just make this goal as well.

The unanticipated factors working against us are:

> 1. We are currently driving at a 25,000 mile a year rate for our two BMW i3's; we anticipated 20,000 miles. These 5000 extra miles are the equivalent of 1200 kWhs.

> 2. We invited a Rotary Youth Exchange student from France into our home for the year--lots of blow drying, curling iron and electronic gizmos.

Part of reason for our journey was to reduce our carbon footprint. We have reduced our carbon emissions to approximately 2 tons for our household per the EPA calculator. This is only 5% of the average household.

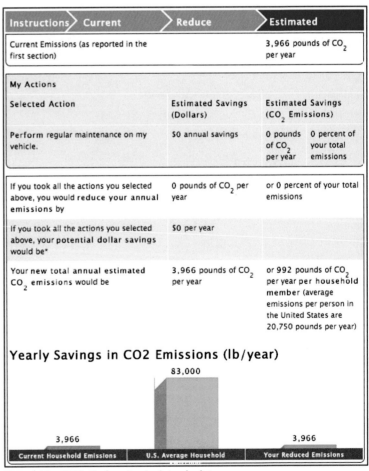

Figure 45 – EPA's Carbon Footprint calculator.

The calculator is very basic and just considers the household and personal transportation. It does not cover air travel or food or consumer purchases, so our actual carbon production is higher.

If we produce carbon, how do we mitigate or offset our carbon emissions?

Figure 46 - Donated system offsets the carbon we produce.

Our choice was not to purchase carbon credits, but rather to donate to our local Agua Hedionda Lagoon Foundation a brand new, 9 kW Solar PV system. This system will produce 14,000 kWh of electricity a year for over 25 years, eliminating between 7 and 10 tons of CO_2 annually. This more than offsets the 2 tons we produce each year and helps a great local cause.

Our yearlong journey "Driving to Net Zero Energy" has come to an end. Thank you to BMWi, ChargePoint, Inside EV's, Stellar Solar, the Center for Sustainable Energy (CSE) and SDG&E for following us and supporting us during this past year.

Our two BMW i3's and our home were powered by harvested sunshine from a small portion of our roof, with zero gasoline cost and zero utility cost for the last year...for forever. We did it!

We believe we are the first household in the world to do this with full documentation. We also believe that there are dozens, perhaps hundreds, of other households who have accomplished the exact same thing without going through the painstaking detail and very public process of sharing private and personal information.

Our year in review:

Let's start with a couple of 800 lb. gorillas and get them out of the way.

1. Although we have no cost of energy or gasoline for our home and two cars, saving us about $9000 a year, we are not 100% energy independent. We are grid connected; we have had a very slight amount of electricity use for the year and a small amount of natural gas use. There is no economic reason to go any farther than below zero cost, but half our goal is to be energy neutral. This has still not been reached. Our costs are zero, and our usage is de minimis.

2. If you think you can plan a year in the future, think again. Our year has unfolded with several unexpected events not foreseen during our planning for the year.

We made a last second decision to host a Rotary International French exchange student for a year so that two young students--one from France staying with us and one from Coronado going to France--could have the experience of living a year abroad. Another person in the household with a full head of hair and an energy hog hair dryer added to our annual total. I actually calculated her monthly energy consumption: 85 kWh. I know: that's a little obsessive-compulsive.

We expected to drive 20,000 miles this year; we ended up driving 21,477 miles.

We lost a family member, priorities about energy savings we're inconsequential during these months of care giving and grieving.

Even as a two EV family, once or twice a year we would borrow a gasoline car for a long driving trip. We take the occasional plane flight, cruise ship, train, bus, and taxi, all of which consume gasoline, avgas or diesel. So, we possess no "EV purity card" and never have, as the rest of the transportation network is not quite as advanced as our household.

We realize how fortunate we are in life and that we are not the statistical norm. However, in California at least, Solar PV is making its way to all income levels and housing types, and EV's can be leased or purchased for well under $200 a month. Living Net Zero does not need to be done in a custom home with BMW's in the garage.

Lastly, we think that's what makes this "Driving to Net Zero" challenge so special. It's a real home with real people, real lives that are wonderfully unpredictably complicated, covered over the course of a year. It's not an unoccupied home on a University campus. It's not a theoretical calculation of energy use by a home builder or a University. It's real, it's us, "warts and all" for a year.

Norby Home & Two BMW i3's Energy Graph

MONTH	JULIE Miles/kWh	PEDER Miles/kWh	SOLAR kWh	SDG&E kWh	SDG&E Cost	Gasoline Cost
May/June	1046/228	1335/348	1432	-24	-$149.20	0
June/July	1011/231	655/150	1353	-118	-$162.57	0
July/August	1385/261	1035/217	1326	-25	-$156.46	0
August/Sept.	1519/331	1342/274	1435	117	-$131.89	0
Sept./Oct.	680/157	677/178	1148	62	-$100.41	0
Oct./Nov.	961/255	623/172	964	218	-$ 13.98	0
Nov./Dec.	485/122	451/115	647	216	$ 40.07	0
Dec./Jan.	862/183	1205/307	710	403	$ 84.73	0
Jan./Feb.	918/236	792/199	905	231	$ 44.47	0
Feb./March	887/194	487/119	1092	-56	-$ 15.48	0
March/April	1136/276	656/177	1293	-235	-$ 53.61	0
April/May	927/235	358/86	1241	-289	-$137.25	0
Cumulative	11862/2717	9615/2344	13546	500	-$751.58	0

Figure 47 - Julie is a more efficient driver, and a credit for the year of -$751.

Now our year of data in review:

Our credit for electricity, $-751

Our annual gasoline, 21,477 miles, $0

Our annual natural gas cost, $280

Our annual total cost of energy $-471

Our energy challenge began on May 15th, 2014, when we received both BMW i3's. Our annual true-up bill from our utility begins every year on January 15th. As you can see we are on a great trajectory for 2015, and we expect this to continue, resulting in an overproduction of around 1200 kWh of electricity. The stronger sunshine summer months are approaching.

Figure 48 - Solar panels occupy about 25% of the roof (in blue).

A total of 5061 kWh of electricity was used to drive 21,477 miles in two BMW i3's.

Julie averages 4.3 miles per kWh

Peder averages 4.1 miles per kWh

We calculated our expected lifetime savings of solar and driving electric, and it adds up to some serious money.

This was calculated when gas in our region was $3.50 a gallon. Today's gas prices are $3.93 a gallon.

Now you see how I can afford to drive the BMW i8 I will tell you about in the next section.

Our 8.5 kW Solar PV system generated 13,546 kWh for the year. This equals 1593 kWh per kW system size.

A 3.18 kW Solar PV system (3.18 times 1593) would provide the 5061 kWh of power for both BMW i3s driven 21,477 miles.

The total cost of the 3.18 kW PV System is equal to buying gasoline for 2.82 years, a ROI of 35.4%.

Two Gasoline Cars and a Home Fuel Cost

Two Gasoline Cars
20,000 miles driven per year
26 mpg cars @ $3.50 per gallon

First year 769 gals of fuel $2,691

25 years 19,225 gals of fuel $67,287 net present cost

25 years, 3.5% annual increase $143,000 fuel cost

Total Fuel Cost $ $143,000

Home Electricity

8000kwh @$0.25 per kwh $2000 per year

25 years $50,000 net present value

25 years, 3.5% annual increase $106,000

Total Electricity cost $ 106,000

25 year Two Gasoline Cars and Home Fuel Cost $249,000

Two BMW i3's & Home Solar Fuel Cost

Two BMW i3's
20,000 miles driven per year
4.2 m/kwh =4762kwh

4kw Solar PV system =6000kwh

First year cost, $12,000

25 years cost, $12,000
Sun has no annual increase

Total Fuel Cost $ $12,000

Home Electricity

8000kwh used per year

6kw Solar PV system =9000kwh first year cost $18,000

25 years cost $18,000

Total Electricity cost $ $18,000

(*System cost and kwh cost are based on a San Diego County location. Solar PV system with micro inverters*)

25 year Two BMW i3's and Home Fuel Cost $30,000

Figure 49 - Lifetime savings calculation.

Driving the BMW i3's powered by Solar PV, the cost per mile is $0.017 cents per mile. If advantageous time of use rates are factored in--as in our case--the Solar PV system size and cost would be reduced by 22%, to $0.013 per mile.

Driving on Solar PV-supplied electricity is 1/10th the cost of driving on gasoline. Overall, we considered the year to be a success!

My Indulgence the BMW I8 - Before I attempt to rationalize the decision, let me say there is no rational reason. It's purely emotional, sentimental, and the fulfillment of a dream. But because of all the previously-mentioned energy and cost savings, it is a dream that became possible.

Figure 50 - I have always loved gull wing doors.

As a young car-loving boy, I sword fought with sections of Hot Wheels track, My Matchbox cars had gull wing doors. I built cars out of Lego bricks that had gull wing doors and big rear wings. Slot cars were great...especially if they had gull wing doors.

The BMW i8 is absolutely expensive, but what price is worth living a priceless dream?

There is an interesting story here, one that I think is worthy of telling, of why there is such a strong attachment and compulsion for me and others to drive EVs and PHEVs.

We're called Electronauts, Fan Boys, Leafers, Voltheads, Teslamaniacs and worse. We are fiercely loyal to our cars, brands, and electric mobility. We EVangelize obsessively and appreciate all supporters of electric mobility. Why so much passion and devotion?

I won't do this justice, but let me try...

One of the greatest needs we have as humans is diverse, meaningful human relationships Money pales in comparison to the value of meaningful relationships.

No greater tool than the automobile has existed for us to connect with each other and to explore the freedom that automobiles provide.. Pictures, texting, and social networking are great, but seeing a distant dear friend in person or sharing a trip to the Grand Canyon with family are experiences that can never be replicated by a photo, text, or phone call. It's such a strong connection, and so important to us, that in many cities we are willing to turn over 50% of our land mass to a mechanical device that allows us to connect face to face.

But there is a huge downside: pollution, congestion, climate change, world conflict over the scarce resource needed to fuel the car, and so forth. Car companies heretofore have been suppliers of both freedom and, increasingly, suppliers of problems.

I profoundly believe that today, car companies like BMW--driven by citizens and regulators, as well as by innovation--are now transitioning and becoming suppliers of freedom and suppliers of answers that will lead us to a cleaner, healthier, more just and equitable world: a world where the car has zero emissions, where the fuel for the car is zero carbon and made on the rooftop, where

Figure 51 - Traffic = Smog.

family and civic budgets are strengthened by the lower cost of fuel and utilities. This will be a world where the air and water are cleaner, where resources are plentiful and equitable, where multiple modes of transit are possible, and where more land is set aside for people and human relationships, not devices.

This is what I desire and what I work for. This is why I have such passion and value alignment for BMW and their efforts.

Of course, BMW needs to manage this transition from producing a few million gasoline cars annually to producing more electric cars, and that does not happen with the flip of the switch. But the transition is unmistakably happening, and BMW is seriously investing in and planning for the electric future. The go-pedal is firmly planted on the CFRP firewall of electric mobility.

The BMW i8 is a master's work of industrial artistry, but is it a Supercar?

By "Old Testament" petrol metrics, the answer is no. However, a quick look at the supercar wiki shows that the definition of what is a Supercar has been in constant evolution.

In the evolution into electric mobility, a Supercar is defined by many as a car that breaks existing boundaries, and is limited to just a tiny fraction of a percent of the cars on the

Figure 52 - The BMW I8 a plug-in Super Car.

road. A Supercar is a car that represents the future in the present. A car that inspires and captivates with a rare mixture of awe and inquiry wherever it travels, offers exceptional rare beauty, sustainability, materials, performance, and road handling and is timeless and treasured for generations.

In the "New Testament," the BMW i8 is most certainly a Supercar.

In January of 2015, the first model year 2014 BMW i8 Electronaut Edition found its way into our garage. The question was: Do I drive the i8 or will it be a garage queen? If I drive it, I could be depriving myself of a good chunk of the monetary value when, not if, it becomes a highly prized collectible. If it's a garage queen stored for posterity and profit, I could be depriving myself of several years of glorious experiences on the open road when my time on earth is done.

I came down on the side of driving it. The plan is to drive our BMW i8 for five or six years--roughly 40k or 50k miles--fix any battle scars, and then put it up on a lift to be used only for that rarest of weekend or road trips.

Supercars are masculine. They're powerful, angular, snarling--some seemingly bordering on evil. After driving and living with the i8, I find the i8 to be a mixture of mostly feminine qualities: sexy, sculptured, goddess-like, exotic, and hard to understand, with many personalities. Perhaps it's the swan-like wings of the opened door, or the flying C pillar reminiscent of a model's long, flowing hair.

The juxtaposed allure of the BMW i8's feminine traits and diverse personalities is why, in my opinion, the car is so spellbinding to so many.

The i8 has many driving personalities--one moment, stealthily, powerfully slow like a Puma without sound, stalking its prey. One moment it's incredibly thrifty and conservative, briskly able to multitask like a coupon-cutting, store-hopping shopper. Another moment it's fierce and fast, hyper-aggressive, poised, lunging, and running at full speed--poetry of form, sound and speed.

Of the three main driving modes, my favorite is slow city cruising in stealth mode to a restaurant or hotel. When button pressing into stealth mode, I always make the swoosh sound as the car becomes silent, moving my hand palm down away from my body. Cracks me up that I do that, but it's the closest thing this side of Tatooine to piloting the Millennium Falcon. The Kessel run in less than 12 parsecs seems doable in the BWM i8.

I'm a 54-year-old "big guy" at 6'3"and 325 pounds. After an awkward first try or two (it's awkward for everyone), I have no problems getting in or out of the i8. With its long doors, it's very easy to swing the legs in and out and the seating position once in the car is perfection. To get out, it's an easy one leg out, pop up and exit the car to the rear so as not to bang the head on the door which is now overhead. Surprisingly, it's one of the easier cars for me to get in and out of.

Figure 53 - The I8 interior, about 40% of my miles are pure electric.

Servicing the i8 in the future should be lower cost than its competitive class. The electric bits, somewhat detuned (ironic, I know) from the BMW i3, will require little to no maintenance, and the 3-cylinder engine from the Mini should be easy to service. Remember that the engine has only 9000 miles on it; for the other 5000 miles, it has just been a passenger along for the ride. Tires are affordable, and brakes are longer lasting due to regenerative braking.

Our driving patterns with the i8 are primarily in two modes. We drive the i8 locally around San Diego North County in all-electric mode. Most of our favorite places are within 5 to 10 miles or less and, with a 15-mile real world electric range, this covers many of our trips. We plug into L2 charging every time the car is in the garage. Approximately 5,000 miles of driving has been in this electric mode, averaging 90 MPGe.

Our daily work driving, and mid-range trips are typically done in our BMW i3s or riding an electric bike.

The other mode is long-distance driving in comfort mode. The i8 is a beautiful and comfortable grand touring car; it's like flying on the road in a Gulf Stream jet. There is plenty of room for gear for two between the rear storage area and the back seat. We've done a four-state, 2000-mile tour though

Arizona, New Mexico, Nevada and California, a few 1500-mile trips to Napa, and a trip or two to Arizona and points east. Approximately 9,000 miles of driving with no plugging in has been in this mode. Where the i8 gas engine averages 30 mpg.

Combined, we average 48 mpg. That's a stunning number for a car in this class. In the near future by shifting to higher density batteries, I'm sure this number will continue to go up as the electric range increases.

Is the BMW i8 a track car? No, it's a grand touring car. The BMW i8 is shy a few hundred horsepower, needs more lateral grip and a stiffer suspension. It can keep pace with the M3 and is just a tad slower than the improved M4 around the track. That's not a criticism of the BMW i8 because it's not set up as an M or track car. When I say that the BMW i8 could use more horsepower and grip, what I'm really saying is that there is so much more room at the untapped top end for this car.

I'd love to track this car with a stiffer and lower suspension, meaty rubber, carbon-ceramic brakes, and a few hundred-extra horsepower on both axels. I'm quite sure the men and women of M could considerably up the ante in a matter of months if given the chance to do so. The BMW i8 chassis and architecture is simply that good.

In a life well led with priorities met, there is room for indulgences So it is with the 2014 Electronaut Edition BMW i8 in our garage. I remember seeing the BMW i8 concept in person at the 2011 LA Auto Show. It was a dream car, something unreal.

Just 5 short years from the prototype 2009 BMW Mini-E to the 2014 BMW i8, I'm living in my own dream. How cool is that?

In Conclusion - There's something traditional, in an American sense, about a home and two cars in the garage. We are a nation that came of age during the past 100 years concurrent with the era of the automobile. For better or worse, our homes and cars are entwined with the embodied energy of our history. We can now power both our homes and our cars with renewable energy, in our case with sunshine. We can also effect change on this personal level of our own lives, although it's harder to do on a citywide or national scale.

The inertia of the status quo is a powerful foe of progress. There is safety and security in the status quo but there is no future, only a past.

It is for this reason that we decided to do the Driving to Net Zero Energy challenge, why I write, why I share as broadly as possible our personal and private information, why I serve as a Planning Commissioner of the nation's 5th largest county by population, so that I--that we collectively--can help break the broken status quo and move ourselves, our families, and our cities to a far better, richer, healthier future.

The gas station down the street, the power-plant in the next city over, and the oil wells across vast seas and deserts, have served our households and transportation energy needs during the past 100 years. They are now giving way to the homeowners who fuel their own homes and cars with electricity from solar harvested from their rooftops.

As a society, we have arrived at a destination, an intersection of our historic calendar and our always-advancing technology prowess whereby today our cities, our homes and our electric transportation choices can be powered by renewable energy generated by a homeowner. During the next 20 years, our country and our individual households will see changes in energy and transportation that are hard to comprehend in their magnitude and benefits to our environment, to our safety and health, and to our household budgets.

We are demonstrating that it is possible, practical, and economical--using today's readily available technology--to construct an energy efficient house, to drive amazing and practical electric cars, and to provide the energy for the house and the cars in the garage via solar PV, all while improving the quality of life and economic situation of the family.

One of the unexpected joys--and I am so fulfilled when I hear it--is the dozens of times in the past year folks have thanked me for writing and for inspiring them and giving them confidence to take steps similar to what we have done.

Just take a small step: a small reduction in energy use or a slightly more efficient car or transportation choice if you can. You don't need to go all "Full Monty" to a make a meaningful contribution to improving your city and your family's budget.

It has been an iterative process for us the last several years, --beginning with solar in 2007 and with two gasoline cars and a toe in the water with the BMW Mini-E in 2009--to where we are today.

Get started and change your world. You **can** live and drive on Sunshine.

Chapter 5 - Radical Practicing Personal Sustainabilist   by Steve Stevens

I was growing veggies as a 5-year-old. Santa brought me a bike 6 months later. I still ride and garden! I also grow grapes, apples, pears, oranges, lemons, grapefruit, and avocados in my 5900 ft elevation home at the 40th parallel in Northern Colorado. I have ridden over 190,000 miles on bicycles, crossing the USA twice and Europe many times. I have ridden to the DMZ in Korea on an 1880s Penny Farthing Bike, before being turned back by US troops, and around the USSR, as well as South Africa. I commuted

Figure 54 - Western Illinois - The rain ended on the great cross-country ride.

to the office in Beijing by bicycle on June 4th, 1990 (the first anniversary of the Tiananmen Square Massacre) riding through the heavily fortified square. I have lived in Colonial Nigeria (1960 – integrating an all-black college) and in Korea.

The antique Penny Farthing is, to me, a symbol for sustainability: demonstrating reuse, repurposing, recycling, and energy efficiency. During my travels to over 60 countries, I often took it for Sunday rides to see the diversity of housing styles and the natural beauty of the countryside. My extensive experience riding this wonderful bicycle gave me an idea. After returning to the States, I decided to convert my 1979 frame home in Northern Colorado to be both my home, and a Victorian Bicycle Museum/ Gallery/ Library – a facility I hoped to design to be as efficient as these old bikes and move beyond net zero!

I am passionate about moving beyond net zero. I lecture about it and teach others the why and the wherefore of cleaning up our world – by cleaning up our own habits and expectations. Camping as a Cub Scout and then an Eagle Scout, I developed a love of nature. I realized that our environment is the only one we have, and our actions can either protect it or destroy it. As a grandfather, I try to plan and act responsibly to protect the generations that follow.

Quick Facts:
> **Net Energy Emissions** – (-7,000 lbs.) - 130% less than the typical American household.
> **Home** – Traditional 1640 sq. ft. Ranch in Golden, Colorado, USA on a walkout basement before the addition of 1300 sq. ft. of "un-conditioned" Passive Solar enveloping spaces built as 13 functional rooms, from large to small, some acting as "airlock entrances" or enclosed walkways.
> **Technologies Employed** – Extreme super insulation, solar PV, electric vehicles, rain water collection, permaculture, passive solar, day lighting, building envelope, solar envelope, air lock entries.
> **Annual Energy Costs** – $ 84, for electrical connection fees.

The road to net zero and beyond began with the creation of the Golden Oldy Cyclery and Sustainability Museum, which included the reconstruction of our home and the construction of the museum space attached to it. Let's first look at the concept of the museum and its goals.

The museum was formed in 2000 as "Golden Oldy Cyclery." The Golden Oldy Self-Challenge/Exhibit had three objectives:

A) To teach sustainability with economically viable and replicable examples in a living museum.
B) To set an example for other museums to publicly convert to a "do no harm" operation.
C) To apply the sustainability challenge to as many operations as possible, including staff.

The Elaborated Challenge:

1) Convert the museum, galleries, and library to be carbon negative.
2) Power the museum complex from on-site, non-combustion-based sources (sunshine).
3) Maintain winter comfort levels (Minimum: 65 / 62° F Day/Night).
4) Maintain summer comfort levels (Never over 75° F).
5) House the staff on-site (no separate carbon footprints).
6) Power the staff's electric vehicles from energy harvested on-site.
7) Feed the staff from permaculture gardens/greenhouse on the museum grounds.
8) Run a surplus of site-produced energy to feed back into the grid.

The Golden Oldy Mission is:

1) To celebrate the glorious history, efficiency, technology, photography, literature, art, and poetry of the world's most sustainable transportation... the bicycle.
2) To do so in the most sustainable museum possible, sharing the methods and motivations with the world.
3) To inspire other museums, as well as other institutions in society, to actively engage their bully pulpits, transforming them to become "Action pulpits" publicly sharing their efforts & results.

Figure 55 - The Golden Oldy Museum is in beautiful Golden Colorado.

The museum collections fall into four main categories:

- Over 60 Pre-1900 bicycles (41 Penny Farthings) and two rare 1876 Ladies High Wheel Tricycles.
- Extensive Victorian bicycle memorabilia and accessories.
- A 120,000-page library of pre-1900 published bicycle journals and 280 historic bicycle books.
- Galleries including 260 framed pre-1900 (primarily Colorado) cycling images, day lit by Sun Tunnels.

Figure 56 - 1876 Ladies High Wheel Tricycle.

It is axiomatic that "The bicycle is the most sustainable of all forms of human transportation." It is only appropriate that a Victorian bicycle museum lead in the Museum Sustainability Movement, to set a new "Do No Harm" standard with transparency of process and results. The integration of the bicycle and climate issues is demonstrated with the Santa bicycle exhibit seen when entering the new museum entrance, through the second passive solar addition. (Santa reputedly did his Christmas Eve deliveries by bicycle before Orville sent, at Santa's request, Wilber Wright up to teach the reindeer how to fly in 1904).

Housed in a 1970s residence, the Golden Oldy structure was built to the period's poor energy efficiency standards. The energy goal for this new permanent exhibit was to go below "Net Zero" energy consumption, while embracing the latest energy needs: electric vehicle transportation and food production. The target audience encompasses both the museum patrons and the broader museum community.

Figure 57 - First exhibit, Santa pleading us not to melt the North Pole.

This latter audience is important as museums can potentially act as a megaphone spreading the message through many "action pulpits" worldwide, showing that actions speak louder than words.

Figure 58 - Some of the many vintage bicycles on display.

The museum building itself is the new exhibit. While Golden Oldy Cyclery and Sustainability has its roots in bicycle history, it now has a broader mission spawning a new permanent exhibit which projects a message for the times. It cuts energy utility usage to negative numbers and cuts carbon footprints to below zero, thus providing carbon sequestration, and it requires no new galleries for its containment … the new gallery structures, as well as updated old gallery structures, are the exhibit.

Marshall McLuhan is famous for his line: "The medium is the message." I have tried to incorporate this concept into the buildings and grounds that are both my home and The Golden Oldy Museum. The building is designed to be below net 0 by producing energy and shipping surplus power to the grid.

Happily, many museums have started on the trail to sustainability. In Colorado, the Denver Zoo, Denver Botanic Gardens, and the Denver Museum of Nature and Science (DMNS) have begun important efforts. This trail to sustainability is not short. It takes perseverance and commitment to pursue the goal. The Golden Oldy intention is to elevate the bar for our eco-circumstances, reflecting, in the words of the late Rev. Dr. Martin Luther King Jr., "the fierce urgency of now."

Figure 59 - Some of the Penny Farthings on display on the lower level of the house.

We are past a point where we can afford to project the message "Do as I say, not as I do."

Museums are constrained to meet their missions while operating within budget and doing no damage. It is not traditional to consider a museum as a perpetrator of damage or harm. However, the built environment has been credited with producing 70% of greenhouse gases globally. Museums have the chance to go beyond being Thought Leaders to be Action Leaders setting very visible & lasting operational examples of good. "'There is no more neutrality in the world. You need to be part of the solution, or you're going to be part of the problem." …. Eldridge Cleaver.

The basic mission of all museums is to educate, inform, entertain, protect, preserve, and prepare the community for the future. Implicit in "PROTECT" is "_To Do No Harm,_" including climate harm. Sadly, traditional museum buildings often represent the worst in eco-operation. Even the best LEED Platinum and Energy Star certifications achieve only approximately 25% to 50% of the goals which are needed to meet today's environmental mandate. In my view, all museum buildings could and should become "eco-exhibits," providing tools for presenting sustainability messages.

The Golden Oldy Cyclery has attempted to meet this challenge. It has earned international recognition in the antique cycling world. In the wider press, it received a mention in the Oct/Nov 2012 **Mother Earth News** article "8 Great Places You've (Maybe) Never Heard Of." Three times it has been the subject of local network-affiliated TV coverage on site and has provided "color content" off-site with static and dynamic displays for many more TV broadcasts.

The key issue of our time is climate change. This is, in the words of Al Gore, an "inconvenient truth."

We are currently experiencing a "slow apocalypse." Left unchecked, the coming natural holocaust will make all past holocausts pale. It will be non-sectarian. However, it will be economically and geographically discriminatory. This threat, unlike the 9-11 attack, and wars based upon nationalism, religion, sectionalism, ethnicity, or simple ego/greed--comes slowly and acts continuously, like the proverbial frog's warming pot on the stove. Its urgency cannot be seen from inside the system without careful studies of volumes of data. Hence, there are many who deny its existence. Western civilizations, and America particularly, grew up with a wealth of fossil fuels powering their development. That abundance created ongoing business-as-usual expectations, without any comprehension of the risks involved. This blind deferral of consequences made short term planning seem profitable, while the frog's pot slowly warms.

The grand threat is that climate change will devastate our economy, society, and citizenry with droughts, floods, super-severe storms, and rising, acidifying seas. The current state of atmospheric degradation is extreme. Greenhouse gas concentrations have grown from the 280 PPM CO_2 level of the Holocene to the current 410 PPM. If not reversed, this concentration will raise temperatures far more than 2°C, causing dramatic melting of the Greenland and Antarctic ice sheets, which will flood vast coastal areas. At the same time, desertification is spreading, reducing food production for the growing world populations. Famine and massive civil unrest cannot fail to follow. Syria has already demonstrated that result.

2012 brought Super Storm Sandy, extreme Colorado wild fires, major crop failures, America's highest temperatures, and dramatically rising food prices. 2013 brought record flooding in Colorado all along the 'Front Range' from storms, which dropped a year's worth of rain in days. The next few years have seen even greater fires, spreading from mountain forests into suburban areas. 2017 demonstrated the increasing power of storms with hurricanes Harvey, Irma, and Maria. The recent California wildfires set new records for scope and damage, some even occurring outside the normal fire season. 2018 has started out with Colorado wildfires which threaten to displace 2012 in the record books. These are just examples of the consequences of climate change. In the words of John Swigert from the Apollo 13 movie "Houston, we have a problem."

The unfortunate reality is that even if all new business, industrial, cultural, and residential construction achieved Net Zero Carbon or PassivHaus standards today, it would not solve the problem. We need to

Figure 60 - Gallery lit by Velux Sun Tunnels during the day.

convert the existing building stock to near Net Zero to stop perpetuating and extending atmospheric damage. We also need to develop atmospheric carbon extraction and sequestration systems to reduce the CO_2 equivalent (CO_2 and CH_4) to well below 350 from the current ~ 490 PPM.

With the broadened scope of the museum to teach sustainability, the collections now include sample cross-sections of super-insulated walls, sample cores of energy exchange ventilators, photo documentation of the step-by-step conversion of the museum, journals, and books on climate, energy, & efficient structures, and displays on the history of oil over the last 15 decades.

This is in addition to displays devoted to the bicycle, especially antique bicycles. The museum's slogan is "The world is a happier place when you ride a Penny Farthing bicycle!"

The House Transformation ... Turning an energy hog into an energy exporter

The homesite is in Golden, Colorado at 5900 ft elevation, with typical summer high temperatures exceeding 100° F and winter temperatures often below 0°F, and occasionally dropping below -20°F.

The core structure was built as a "2 by 4" frame, three-bedroom ranch home, sited on a 0.19-acre lot. The windows were a leaky aluminum-framed, non-low-E variety. The house included a walkout basement whose large family room now provides the primary bicycle display area.

Figure 61 - Before the transformation, a typical 70's 3 bdrm Ranch with 2 Baths, high energy bills, and uneven heating, resulting in uncomfortably chilly north rooms.

The basic approach to our energy renovation is to catch the available solar and retain the energy in an airtight, well-insulated building envelope. We added four unheated additions that protect and buffer the core house and provide many useful spaces, airlocks, and rooms, as numbered below.

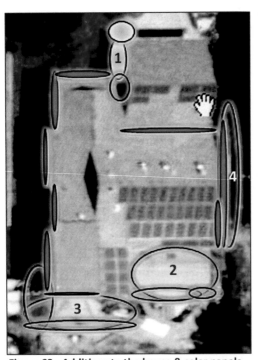

Figure 62 - Additions to the house & solar panels

To the left is an overhead view of the home/museum. Red ovals indicate where we super-insulated the outside walls. We could not add an envelope due to bedroom egress codes. Blue ovals indicate envelope additions to buffer, insulate, and act as an airlock or sun catch space. Currently, all entries have a double airlock, except the front door, which has a single airlock. This will be addressed with a future outer airlock (north of # 1). These additions help protect and insulate the core house/museum, while providing useful space. Also seen in this view are the solar panels used to make our power. The solar panels are not visible from the roads as they are on the back side of the roof pitch. The first addition (**1**) was a small front airlock vestibule on the north side. The second addition (**2**) has the white-roofed, two-story space on the south side. It incorporates a 25' by 10.5' conference room with an outer glass-enclosed 2.5' by 19' walk-in solar clothes dryer and an airlock entry. These spaces are atop two workshop areas. This addition has 214 sq. ft. of passive solar glazing to the south. In summer, eve overhangs shade the south glass. The third addition to the south-west (**3**), created a triple-walled passive solar envelope complex. On the main level is a 6' by 20' gymnasium, wrapped with a 1' by 20' heat collector which directs heat upward, better shown in Figure 65. Below the gymnasium is a greenhouse, and above is a passive solar plenum that acts as a passive solar food dehydrator. The main and lower levels have airlock entries on their northwest sides. South-facing solar glazing totals 426 sq. ft. Heat gained from these spaces is piped to the north end of the house through three insulated 12" air ducts. In summer, these spaces are thermally vented with four windows at the top level and two at the mid-level. The fourth addition (**4**) is a 40 ft by 5 ft east side enclosed walkway. It has 64 sq. ft. of south-facing solar glazing. It encloses the PV inverter and electrical equipment (previously outside) and thermally buffers the home, better shown in Figures 71 and 72.

"Passive" energy strategies include both passive solar (solar space heating) and PassivHaus (super-sealing and super-insulation). The specifics include:

- An aggressive air-sealing effort to stop convective losses.
- Ceiling insulation was increased to at least R70 and, where space permitted, to R150 by filling the attic spaces up to the rafters with cellulose--recycled and fire-proofed newspaper.
- Windows were replaced with Low-E triple-glazed units supplemented with cellular window shades.
- Gas heating was upgraded to a Fujitsu mini-split heat pump.
- An Energy Recovery Ventilator (ERV) provides healthy air qualities, constantly exchanging the indoor air with fresh outside air while transferring the heat from the stale outgoing air to the incoming fresh air.
- Daylighting features include seven large Sun Tunnels (22 & 14 inch), two instances of glazed flooring for inter-level light sharing, and light-colored floors/ceilings/roofs.
- We added a 10 kW rooftop PV system producing more energy than we use.
- We dug around the foundation to add 4" of poly-iso foam insulation.

Figure 63 - The dining room wall was initially upgraded to R-68 with inside foam, then to R-104 with outside foam. It is now 22" thick, with 13" of Poly-Iso rigid foam.

Figure 64 – Five and a half inches of foam insulation was added to the east exterior, bringing R values in the kitchen to R-45 and to R-104 in the dining room. Recently, this area was enclosed in a 40' X 5' space providing two climate zones of protection as well as an enclosed walkway down the eastern side.

- A 10" glazed envelope space on most of the south side of addition (**2**) --the gymnasium and conference rooms-- captures the sun's heat in winter. This heat is ducted to the north side of the home as needed. In the summer we duct the warm air outside. This passive solar layer complements the insulation and sealing efforts.
- The overhangs help reduce summer heat gain along with thermal siphon venting by opening the outer windows.
- We transitioned to plug-in cars to use excess PV production to power them with clean energy.
- Winter sunshine is gathered in addition (**3**) shown in Figure 65. These rooms were added using mostly recycled, surplus, and otherwise inexpensive, but quality, materials. For example, we found a lot of thirty glass, triple-pane freezer doors that made for great glazing. The staff did the planning, scavenging, and drafting of the permits. Then, along with a neighbors' expert help, they did the construction.

Figure 65 - Back side of 3rd addition incorporates a large solar collector.

- All entrances have airlock additions … like Jules Verne's Nautilus and the Space Shuttle. Most entrances have airlocks on airlocks. There are no drafts upon entry or exit to the museum home! Most of the airlocks were designed as passive solar additions.
- The airlock to the greenhouse serves as a potting room for gardens and provides water-based thermal mass for temperature stabilization.
- We added a 'climate battery' of earth tunnels in the greenhouse and water tanks as thermal masses to moderate temperature swings.
- Water heating was converted from natural gas to a solar-powered electric heat pump.
- All electrical lighting has been converted to LED.
- We bermed the foundation on the north side for winter warmth and summer cooling.

Figure 66 - Air lock to greenhouse with water & brick based thermal mass.

Figure 68 - Solar plenum also serves as a food dehydrator. Grapes become raisins, pumpkin slices become pumpkin chips, etc.

Figure 67 - Exercise room and outer solar envelope feed heat up to the solar plenum. This area is above the greenhouse.

Solar spaces provide second functionalities in addition to buffering the core house such as:

- A staff gym (5 recycled machines from Goodwill) located within the inner solar envelope wall.
- The solar plenum also serves as a solar food dehydrator, are part of addition three.
- A museum conference room
- A bicycle restoration workshop
- An indoor dwarf citrus orchard
- An airlock serving as a daily usage bicycle garage
- Solar clothes drying envelope

The home/museum remains an ongoing project; the conference room is now being remodeled with even more insulation.

Figure 69 - Conference room. Notice the double wall on the right and far end.

The south wall second addition, on the site of the original rotten south deck, was made with recycled thermal-pane glass patio door components from Habitat for Humanity's Re-Store to catch the sun. It was good in the day, but at nighttime it lost a lot of heat because of its low R-2 insulation value. I added an R-3 cellular blind shortly after to get R-5. While this was a great nighttime upgrade, it still was clearly a three-season space. Then I added the inner glass R-3 glass wall, separating the room into a conference room and walk-in solar clothes dryer. While this was an improvement, night temps still fade a bit in this unconditioned space. The outer south wall of this first passive solar addition totaled R-8 at this point. I then procured some used R-13 (center of glass) four-layer insulated glazing units (IGUs), which had minor inner wrinkles in the inner mylar areas which can be seen if you are looking at the window

critically, but not if you are looking through the window. I am re-framing the inner wall of the clothes dryer just outside the conference room to accept these 'super windows'.

When done, the total south wall of the conference room will be R-18 and the south wall of the original inner house will be R-27, with a big sun-catcher on the south side. While this is still less than the other walls on the house (R-60 to R-104), it should be a great improvement on cold (below 0°F), gray days.

Figure 70 - Conference room glass wall being upgraded from R 8 to R 18.

Our fourth addition is an east-side walkway. As in other additions, it is not heated or cooled, but acts as a buffer, so the core house/museum is not exposed to the extremes of the weather.

This attractive thermal break and air lock entry, addition (4)- keeps the walkway from front to back ice free, provides storage for tools, and keeps the solar inverters dry.

Figure 72 - The 4th envelope – completed October 2017: An east side enclosed walkway / herbarium / tool storage area.

Figure 71 - Inside view of 4th addition.

The Grounds:

- Permaculture gardens and dwarf orchards replaced most lawn areas, providing 22 vegetable varieties, 12 grape vines, and 8 dwarf fruit trees (with 28 varietal grafts).
- The garden soils (outdoors & inside the greenhouse) were heavily amended with bio char, compost, alpaca poo, and peat moss for natural fertilization.

Figure 73 - West side view of house and outside garden area.

Additionally, photovoltaic panels provide _active_ solar collection with a roof-mounted 10,000-watt array that is grid-tied, but with battery backup. The solar panels are barely visible from the street, as they are hidden behind the roofline. The solar PV more than offsets our usage for the house/museum/library/gallery and both our plug-in cars.

A video tour of the house is available at YouTube: http://www.youtube.com/watch?v=t3jvittg60I

Plug-in Cars

Our first experiment with plug-in cars was to add an extra 5 kWh battery and plug-in charger to our 2007 Prius. This greatly improved the mpg around town, and since we charged from our PV system, it was also cleaner. We then added a second plug-in car, a Chevy Volt. Its 38-mile electric range allowed us to do nearly all our local errands on sunshine.

Figure 74 - Initial 2007 converted Plug-In Prius led to the Chevy Volt. That led to a used Tesla Model S, which was traded in to Tesla for a Model X in 2017. This was an educational evolution of auto electrification.

Later, as we tried to eliminate our gas usage, we upgraded to a used Tesla Model S, which we later traded for a Tesla Model X, to give us needed hauling and utility capabilities.

The Model X 90D has over 250 miles of range, allowing us to do most of our driving cleanly with the electricity we produce on the roof. For longer trips, we use Tesla's Supercharger network, allowing us to drive virtually everywhere in the 48 states and Canada--and parts northern Mexico--with lower emissions, for free. Since TESLA is a solar energy company, they are fueling Superchargers, where possible, with electricity from renewables.

Figure 75 - Tesla Model X: capable of being a camper or material hauler.

Quantitative Results

The results are dramatic in many dimensions:

- Tour requests for the museum's sustainability aspects now exceed the antique bicycle interest.

- Many energy-focused organizations- the National Renewable Energy Laboratory (NREL), Sandia National Laboratories, the Illinois Institute of Technology, and others--have requested and received presentations on the museum's energy transformation. Also, papers have been given at "Solar 2011" in Raleigh, N.C., "World Renewable Energy Forum 2012" (WREF) and many other venues. I was most happy to receive an invitation from the Society of Friends (Quaker Church) in Boulder, Colorado to provide a Sunday morning lecture. I found their ethics and thoughtfulness to be interestingly refreshing.

Figure 76 - Energy Renovation award given by Governor Ritter.

- At WREF, former Colorado Governor Ritter, who originated the term "New Energy Economy" in 2005, presented the Colorado Renewable Energy Society's award for Best Residential Energy Renovation to Golden Oldy Cyclery & Sustainability, with CRES's Executive Director Lorrie McAllister assisting.

After years of incremental efforts, the structure's annual energy and carbon footprint are dramatically reduced, even while supporting power loads for new functions. The following charts show the energy/climate improvements since 2000. Natural gas usage has declined from 1095 therms in 2002 to 0 in 2018. The replacement Fujitsu 12,000 BTU cooling/ 16,000 BTU heating mini-split heat pump has been installed. It has an impressive 29 SEER, about double that of an Energy Star heat pump. This allowed us to cut off the natural gas line, which at $100/year was our only energy cost. We will find a new home for this orphaned, Amana 95.5 percent furnace.

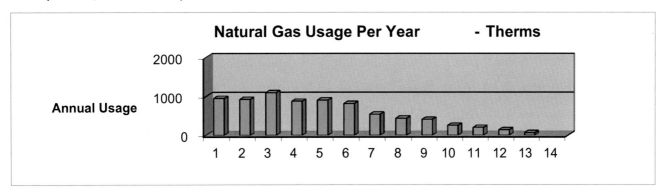

Figure 77 - For 2017, only 31 therms of natural gas were consumed. This is a reduction from 1095 therms in 2002.

In 2012, we over produced ~ 7,500 kWh of electricity, which was shipped to the grid. The recent decline in surplus solar-produced electricity reflects converting the auto fuels to electric and the greenhouse production of citrus fruit. Also, the south neighbor has maturing Blue Spruce trees, providing an ever-increasing shading challenge going forward.

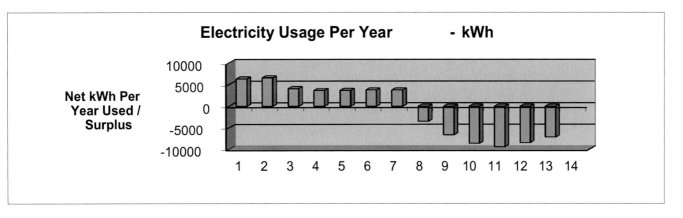

Figure 78 - We still produce a surplus of electricity, even powering the Tesla Model X and Chevy Volt.

Solar Access: The home immediately to the south of our home/museum has seven Blue Spruce and one Silver Maple near the lot line. Although an agreement was signed and recorded in 2007 to allow trimming of the trees to a certain level as they mature, the south neighbor has verbally rescinded the rights. The trees are now slowly, year by year, reducing the solar access of our property. We don't want to create a legal issue at this time, but the sun's gift is being reduced annually. Ancient Greece and Rome, as well as the ancient Navajo Nation, had solar access laws which basically said, "Thou shall not block thy neighbor's winter sunshine." Our modern society is not as wise.

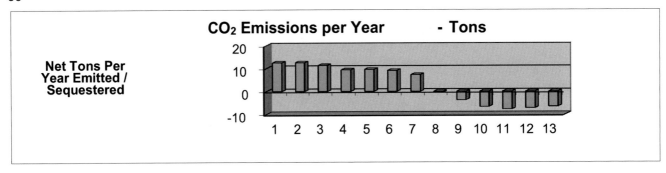

Figure 79 - In 2012, we sequestered over 6.5 tons of CO₂ while powering two plug-in cars.

In 2007 the structure crossed over to the eco-friendly "doing no harm" level, as we used less power than we produced, thus making us carbon negative.

Scaling and Location Issues: Many have told me, "But my building is too big/too small." We reject the contention that size matters. Ours was not the Goldilocks "just right" size. Additionally, we are not in a Goldilocks climate (records: -48°F to 105°F). Finally, we were not able to start with a clean new construction, as we retrofitted a 39-year-old building that was *not built* with energy efficiency in mind. Going to net zero is a matter of awareness, heart, priorities, and intention. To host a museum is a trust and a privilege. In the words of Noam Chomsky, "The more privilege you have, the more opportunity you have. The more opportunity you have, the more responsibility you have."

Conclusion

As Andrew Carnegie said, "Do real and permanent good in this world." In doing that, the Air Force Motto "Aim High" provides guidance. Token efforts are not enough. Robert Kennedy stated, "There are those who look at things the way they are, and ask 'why?' I dream of things that never were, and ask 'Why not?'" Gandhi said, "Be the change which you want to see in the world." None of these approaches are "business as usual." They are not profit driven. They are human- and earth-focused. Yet, they are affordable and yield a good rate of return in energy savings. The greatest rate of return, however, will be in generational savings. Currently in human history, with runaway population, severely depleted natural resources, atmospheric and marine destruction, and major loss of species, we need to adopt the Japanese concept of Satoyama: "living in harmony with nature." Museums have a prominent place in society to lead by providing examples of permanent, intentional, & effective action.

Figure 80 - The upgraded house/museum produces more power than needed for the building and two plug-in cars.

Chapter 6 – The Heliospiti (Sun House)   by Jim Riggins

My wife, Elise, contends that the most dangerous words in our house come when I am looking online at DIYNetwork.com and say, "Well, that doesn't look so hard." Wanting to combine my love of working with my hands, and wanting a sustainable home, I decided, for the first time ever, to make all our new kitchen and bathroom counter tops with concrete and recycled bottle glass. For the record, it looks a lot easier online!

Figure 81 - Making Countertops

Quick Facts:
  Net Energy Emissions – 3000 lbs. - 93% below average household.
  Home – New Net 0 construction, 3200 square feet at 7000' elevation in central Colorado, U.S.
  Technologies – German Passive House design; solar photovoltaics; solar hot water; passive solar; earth tube ventilation & pre-conditioning
  Energy Costs – $0 for house and first electric vehicle; Average $23 per month with 2nd electric vehicle (EV)

Figure 82 - The Net-Zero Heliospiti and our all-electric transportation.

Background - After thirty years of military moves, my wife and I decided to combine our passion for sustainable living, energy efficiency and renewable energy with the construction of our retirement house. Having begun my own home energy rating and consulting business, we felt compelled to practice what we preached when it came to our own house.

For our "Heliospiti" (Greek for "sun house") project, we performed the complete energy design and computer modeling, and in the process of months of research, compiled a set of tools and resources that may save time for others choosing this path.

My passion for sustainability came from my teenage years living through the Arab oil embargo of the early 1970s and the news of environmental disasters, such as the Love Canal tragedy in Niagara Falls, New York, that were decimating nature and killing people. My wife's environmental awareness came from being the daughter of a nature-loving outdoorsman who chose camping, hiking, and national parks over tourist traps for family vacations. We are also both children of folks who lived through the Great Depression and instilled in us the virtue of conservation.

When we retired from the military, we settled on Monument, Colorado to build our retirement home. We found an open lot with good solar exposure within walking distance from hiking trails of the Pike National Forest. The lot did, however, sit at 7000 feet elevation, so we knew we had our work cut out for us to get to net-zero energy in such a heating-dominated climate.

Design Goals - Our goal was to build a pure net zero energy house, emphasizing passive features, while minimizing environmental impact both during and after construction. We defined "pure" net-zero as meaning that we would not emit natural gas or propane exhaust into the atmosphere but offset our electrical fossil fuel energy with solar electricity production. We wanted an all-electric house where we produced 100%, or more, of the consumed electrical energy.

For the more than two decades of constantly moving with the military, I would form design concepts in my head for this home. Articles about geothermal heat pumps, wood-fired hydronic boilers, and masonry heaters convinced me (wrongly) that one builds an efficient house around an ultra-efficient heating and cooling system...and then I learned about the German Passive House philosophy and all the light bulbs went off in my head. Simply stated, the Passive House design philosophy combines a super-insulated and air-tight shell with passive solar design to vastly reduce heating loads. The heating loads should be driven down to such tiny numbers that a very small, conventional, reliable, and inexpensive mechanical heating system can satisfy the peak heating loads. As an engineer, which means I'm obsessed with efficiency and abhor waste, I fell in love with the elegant simplicity of Passive House. I also discovered a U.S. Passive House network, some of whose members assisted me with design decisions and overcoming dilemmas.

A second goal was to power the full house and a yet-to-be-purchased electric car, with the smallest solar photovoltaic (PV) system that would take us to net-zero energy. This meant a relentless quest to count every consumed watt, to use only ENERGY STAR® appliances, solar hot water, high-efficiency lighting, and efficient auxiliary motors for items such as our water well pump and solar hot water distribution.

A third goal was to have extremely low water consumption as compared to a typical American household. This seemed like a moral imperative when living in a high desert region prone to periodic severe drought and where water is a precious resource.

Finally, we wanted to address total sustainability in the house, not just energy and water conservation. This meant maximum use of recycled and recyclable materials in the house. It meant using only non-off-gassing materials and zero or very low volatile organic compound (VOC) glues, paints, stains, and varnishes.

This chapter will take the reader through our journey from initial concept through construction and our assessment after over six years of net-zero energy living.

Where to Start...

The general approach we pursued was to a) select a basic house shape and size that was big enough for our current family of four, but small enough to be our empty nest home; b) an open floor plan with no wasted space in rooms we would hardly use; c) build a baseline reference home in the computer energy model using the current local energy code; d) model various permutations of wall, window, roof and foundation insulation to study performance and tradeoffs; e) after selecting the final shell design, size the solar PV, solar hot water and backup mechanical heating systems.

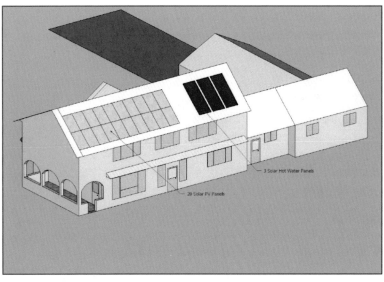

Figure 83 - Initial Conceptual Design Performed in Google SketchUp

We wanted a basic house shape that optimized passive solar: rectangular with one long face facing south and narrow enough to allow deep penetration of the sun during the winter months. We selected a two-story house to provide sufficient south-facing surface area for solar gain, while minimizing the footprint. A slab-on-grade, versus a basement foundation, would give us the thermal mass "heat battery" to store and modulate the passive solar energy.

Rather than start from scratch to design our own floor plan, we searched for an existing plan that matched our requirements. I had read some excellent passive solar articles by architect Debra Rucker Coleman in Home Power Magazine and searched the website of her company Sun Plans[18], where I found the perfect design for our starting point. The floor plan was exactly what we were looking for. All we had to do was convert an excellent passive solar design into a net-zero energy house using the Passive House philosophy! This, of course, was much more complicated than it sounds.

The Energy Design

With the basic house shape, size and orientation determined, I could begin the computer modeling process.

For my building energy consulting business, I had a powerful model, ENERGY-10. Developed by the National Renewable Energy Laboratory, it provided hour-by-hour energy modeling over 365 days. Hourly modeling is critical for high performance homes to capture the contributions of passive solar.

We took a standard design-engineering approach of parametric analysis for our modeling. Simply put, this means using the baseline house but changing only one parameter, wall insulation for example, and analyzing how the heating, cooling, and electrical loads for the whole house change, for varying levels of that one parameter. What we were looking for were the points of diminishing returns. In other words, at what point for wall, window, roof, and foundation insulation does adding one additional R-value have a small effect on overall building energy?

For example, Figure 84 shows the relationship between wall R-value and annual energy consumption. The R-13 left-most data point is for a 2x4 framed wall. The right-most data point of R-120 is just an artificially high point after which the impact of adding more insulation is close to zero.

The graph shows a well-defined knee at R-values between 40 and 50. This is the point of diminishing returns, and what we used to select our desired wall insulation level. The same analysis was applied to the concrete slab and roof insulation. It showed diminishing returns at approximately R-18

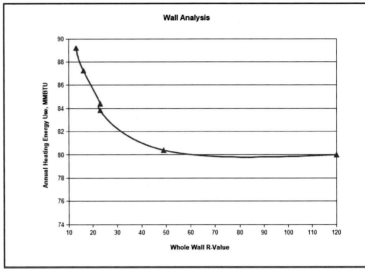

Figure 84 - Parametric Analysis of Varying Wall Types

for the slab insulation and R-65 for the roof. Windows were more complicated to model because two variables can change. The first is the U-factor, which is a measure of the heat transfer through the window. The other is the solar heat gain coefficient, which is a measure of how much solar heat passes through the window, a critical measure for a passive solar home, especially for the south-facing windows. We spent many more days modeling different windows, double and triple pane, using manufacturer data and varying the U-factors and solar heat gain coefficients. Given that the cost of windows varied wildly by over $100 per square foot for different performance and manufacturers, selecting them was one of the more difficult design aspects of the house.

Where all the insulation parameters showed non-linear graphs with well-defined bends, one parameter was quite different. Figure 85 shows the relationship between building air leakage and total annual energy consumption. There is a minimum (zero leakage) value and a maximum value, but the relationship in between is a straight line. This is what makes eliminating air leakage one of the most cost-effective means to a high-performance house. Our goal, then, was to get as close to air tight as we could and then install mechanical ventilation to maintain a high indoor air quality.

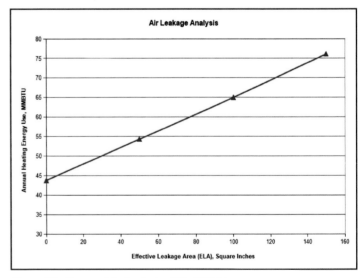

Figure 85 - Parametric Analysis of Shell Tightness

Construction

The main level slab floor serves as our thermal mass to capture the winter sun and release it at night and on cloudy days. It is concrete four inches thick and thermally isolated from the ground and the foundation stem wall by R-20 foam insulation.

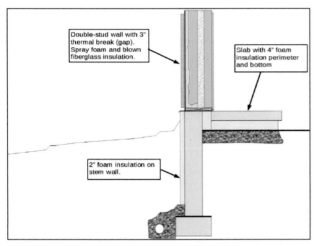

Figure 86 - Foundation construction and insulation details.

Figure 87 - The foam "tub" that insulates the concrete clab

We selected a double-stud wall configuration as the best compromise between cost and performance. The 2x6 outer and 2x4 inner framing are separated by a three-inch gap which serves as a thermal break to keep heat from conducting through the studs. The total wall thickness is twelve inches. A combination of sprayed foam and blown fiberglass creates a very air tight wall with R-49 insulation.

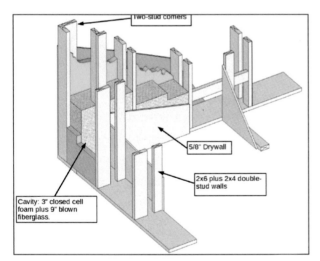

Figure 88 - Double-stud wall construction and insulation details.

Figure 89 - The exterior walls under construction.

This was not as expensive as one would think as the 24" spacing helped offset the cost of the double wall and our HVAC system is much smaller than normal.

The roof is insulated, and air sealed at the deck to allow for the attic to be conditioned and used as a mechanical room for the solar hot water drain-back tank and the energy recovery ventilator and ducts. The trusses are filled with five inches of spray foam and blown fiberglass to give a tight roof with R-67 insulation.

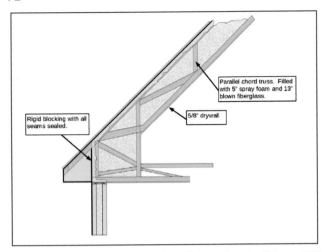

Figure 90 - Roof insulating details, yellow is closed cell foam, R65.

Figure 91 - The roof framing during construction.

We accomplish the passive solar winter heating with triple pane, high solar heat-gain windows on the south face. The east and west faces contain no windows, and the north face contains the minimal glazing for ventilation, day lighting and emergency egress. As seen on the right side of Figure 92, a passive solar wall, which relies on convection only, provides heat to the workshop and garage.

Figure 92 - South-facing windows provide passive solar heating.

Figure 93 - Solar heating of the concrete main floor.

Overhangs on the south-facing windows are designed to block the high-angle summer sun but allow low-angle winter sun to enter the home. These overhangs allow 100% of the sunlight to enter on December 21st (at solar noon), and 0% on June 21st.

The four-inch thick stained concrete slab serves as the main floor and the primary thermal mass for the passive solar. It absorbs the winter sun and will slowly release the heat into the house when the sun is not shining. It smooths out large temperature spikes, in both summer and winter, that would otherwise occur.

Modulating indoor temperatures with the concrete slab and using overhangs to block summer sun are just two of the elements of the passive cooling design in this house. The other element takes advantage of the cool, dry evenings that exist at our elevation in Colorado.

Figure 94 - Stairwell is central to passive cooling.

We located the stairwell to the second floor in the center of the house and left it open to serve as a "thermal chimney" in the summer. By opening low windows on the first floor and high windows on the second floor at night, the hot air rises and exits from the second floor and pulls in cool dry outside air across the concrete floor. This keeps the house cool and comfortable throughout the day, even with the windows closed and outside temperatures greater than 90°F.

Supplemental heat is provided by a single Mitsubishi, 9000 BTU per hour mini-split heat pump. We also installed a small wood stove with a dedicated combustion air duct to the outside. In a tight house, drawing air from inside the house can create a dangerous negative pressure that pulls combustion gasses into the living space.

Figure 95 - Backup heat from small wood stove.

Figure 96 - A Mini-Split HP provides backup heat.

A tight house requires ventilation to expel water vapor, carbon dioxide from people and pets, and fumes that might be emitted from furniture, paints, and cleaning supplies. To avoid the energy penalty of simply exchanging inside (heated) air with clean outside (cold) air, we installed an energy recovery ventilator in our attic. This unit transfers more than 80% of the heat energy from the stale exhaust air to the incoming fresh air. We can turn off the heat exchange function in the summer if we can't open windows and require ventilation. To preheat the winter incoming fresh air and precool the summer air, the inlet to the ventilator is connected to a one-hundred-foot-long earth tube buried ten feet below the surface where the earth is at a

Figure 97 - An earth tube moderates the ventilation air temperature.

Figure 98 - The energy recovery ventilator.

constant 52 degrees F year-round. A filtered, vertical inlet tower stands four feet above the ground to prevent snow blockage.

In our climate zone, water heating is second only to space heating in overall annual energy consumption for a typical family. To overcome this large energy drain, we installed a three-collector solar hot water system. The fluid that circulates to the roof for solar heating then heats our domestic water through a heat exchanger in a 120-gallon storage tank. The tank has an electrical backup element, but we have turned that on only once in seven years. A combination of the large storage tank, ENERGY STAR washer and dishwasher, and low-flow showers and faucets has allowed our family of four to rely only on the sun for our hot water needs.

Figure 99 - Rooftop solar thermal water collectors.

When the heating and water heating loads are driven down to extremely small numbers using efficiency and passive solar, then appliance and lighting loads quickly dominate the total energy use. Our strategy to attack this was to install only ENERGY STAR certified appliances and use high-efficiency lighting. The original lighting was all fluorescent and compact fluorescent, given the very high cost of LED lighting at the time. We have since transitioned the whole house, including tube bulbs and refrigerator bulbs, to LED. Because we draw water from a 600-foot-deep well, water means electricity to us. We economized on water-saving fixtures, and we were able to save significant energy by having a high efficiency, variable speed well pump installed.

After loading every measured, known or estimated electrical consumer into a spreadsheet, we calculated that a 4,400-watt solar PV

Figure 100 - The 4.4 kW solar PV system installed on the back side of the house.

system (20 panels) would be sufficient to power the entire house and have a little excess to offset part of an electric car. The house is "grid-tied," so we use the local electric utility as our battery and avoid the large cost of a battery and management system installed at the house.

Performance

Table 1 below summarizes the measured first-year performance. The tracking software that came with our solar microinverters provided the total solar electrical production figures. The net meter of this all-electric house provided the net usage from the grid. With these two numbers, we could calculate how much energy was produced, how much was consumed, and what was the net total. The table shows that the house exceeded our expectations. Without the electric vehicle, we produced an excess of 3532 kWh for the year. This averages out to an excess of 9.7 kWh per day for the full year.

Category	Estimated Consumption (kWh/yr)	Actual Consumption (kWh/yr)
Appliances & Pumps	3,166	4,053
Lighting	864	(included in above)
B/U Water Heat	680	0
Space Heat	1,550	0
House Total	6,260	4,053
Elec Vehicle (LEAF)	10,512	2,190
Total Consumption	16,772	6,243
Total PV Production		-7585
Net Energy		-1342

Table 1: Heliospiti first year performance. We came out ahead.

During this first year, we only had the electric LEAF for two months. But the data was very promising and suggested that the solar PV system would cover all our vehicle recharging needs in addition to the entire house.

The analysis at the end of year five in the house proved this to be true. Based on the initial meter reading at move in, and the reading on the five-year anniversary in May 2016, the net excess production was 8829 kilowatt-hours (kWh), which averages to 4.8 kWh excess per day. This includes recharging the LEAF for fifty of this sixty-month period.

Eliminating Gasoline

When we started designing our house in 2009, there were no mass market electric vehicles. Toyota and General Motors were abandoning their electric vehicle programs, and the future looked bleak. I had researched performing an electric conversion of a gasoline vehicle and assumed that would be the best we could do. When Nissan announced the LEAF, and its performance specifications became clear in

2011, we were hooked. We purchased one in April 2012 and not only was it well engineered and fun to drive, but it covered 80% to 90% of our daily local personal and work driving.

Replacing our vacation and long-distance travel vehicle, an aging 1993 Chrysler minivan, with an EV, seemed like much more of an impossible dream...until Tesla destroyed the status quo of personal transportation and built EVs with 200+ miles of range, and just as importantly, installed the nationwide charging infrastructure to make all-electric cross-county driving a reality.

In August 2016 we took delivery of a 75 kWh, Tesla Model X, which completed our transition away from gasoline (except for a riding lawnmower for now) but also caused us to pay our first electrical usage bill in seven years. The combination of two EVs and an electric house finally exceeded the total output of our 4.4 kW solar PV system. The average net consumption since the addition of the second EV is approximately 189 kWh or $23 per month. We plan to address this with a future addition of six more solar panels.

The Net-Zero Energy Lifestyle

For homeowners who wish to do nothing except switch a thermostat between "heat" and "cool" twice a year, a net energy house may not be for them. However, after more than seven years in our house, we find the additional workload to be straightforward and minimal.

Winter indoor temperature control on sunny days may require lowering our translucent blinds, or cracking open a window or two, to prevent overheating. We also look at weather forecasts, and if an extended period of bitter cold, snow, or fog is headed our way, we will allow the indoor temperature to climb to 76 degrees F, to capture additional heat in our thermal mass. We like to call this "banking BTUs (British Thermal Units)."

Likewise, if we see extended periods of limited sun in the forecast, we will do all our clothes washing while it is still sunny and limit washing to cold water only during the limited-sun periods. But these issues only apply to periods of four or more days of limited sun, which are rare in Colorado. The thermal and hot water storage easily provide two to three days of back-up if we are receiving very limited solar energy.

When the weather allows it, we dry clothes in the giant solar clothes dryer--outdoors on a line. This not only adds labor to our lifestyle, but sadly, many homeowners' associations around the country prohibit the use of clotheslines. This is unfortunate because dryers operate at very high-power consumption.

Other behavior patterns should be universal, whether one lives in a net-zero house or not. We turn off lights when leaving a room. Whenever possible, we use natural daylighting and not electric lighting. We are very conservative with water use, and the pump electricity it uses, by using only roof-collected water for irrigation. We unplug electronics when not using them to eliminate phantom loads. We combine driving trips to minimize miles and recharging.

Taken altogether, the transition to a net-zero energy lifestyle was quite easy and is now second nature to us. Additionally, it also means keeping a lot of green in our pockets as we pay an average of $23/month to power both our home and two plug-in cars making for a very attractive compromise--a fee we plan to eliminate with a few more solar PV panels.

Conclusion

Even after seven years, the thrill of hearing the solar hot water pumps turning on, or watching our solar electric production output, has not diminished. We like the feeling of knowing that we our powering our home and most of our transportation with free, clean, lung-healthy solar energy and not contributing to health, environment, and climate degradation through the use of fossil fuels.

The added cost of building a net zero energy house was much lower than we expected. All the passive features—insulation, air sealing, high gain windows—added 0.7% to the typical cost of our general contractor. After adding the energy production systems—solar PV, solar hot water, earth tube—the total cost of the house was 8% higher, after state and federal incentives.

To get this message out, we use our house as a community demonstration. We have shown it on the National Solar Tour for two years and given numerous individual and school class tours to provide public education.

While there are a few items we would do differently if building today, they are relatively minor. Our high solar gain, south-facing windows allow more ultraviolet light in the house, which has faded the water-based wood dye on cabinets, wood window sills and trim. I am slowly refinishing these and using a UV blocking exterior polyurethane finish...which I should have called for during construction. Lighting technology and cost have also advanced in the seven years since we move in. In a new house today, 100% of the lighting would be LED. And finally, of course, there is always the second-guessing on the

floor plan. I should have made the mechanical room, which is also the laundry room, a little larger to leave room for a clothes-folding table or shelf. We also plan to add a few more solar panels to bring us back to full Net 0 energy use. But these are minor issues.

Overall, we are thrilled with our net-zero living and electric driving and will never go back. I'm even glad we made our own counter tops which, while more difficult than I expected, turned out great!

Figure 101 - Home made countertops of concrete & glass ended up great.

Chapter 7 - Freedom from Fossil Fuels: An Architect's Quest For a Zero Net Energy Home
by Alan Spector, AIA

As an architect in the 1980's, I made a commitment to work towards a renewable energy society. Since then, my wife, Mary and I have been continually working to implement that goal in our own life by building a passive solar home in New Jersey in 1984, and over time adding a solar hot water system, mini ductless heat pumps, two solar photovoltaic systems, and finally achieving a zero-net energy home in 2016.

What a great feeling to be free from fossil fuels!

QUICK FACTS:
> **Net energy emissions** – 2050 lbs. - 95% less than average, or 100% with carbon offsets.
> **Home** – Passive solar design with 2,751 sq. ft. + 400 sq. ft. in sunspace, northwest New Jersey
> **Technologies Employed** – Passive solar, Solar PV, Solar hot water, Mini-ductless Heat Pumps, Electric Vehicle, Carbon Offset Credits
> **Annual Energy Costs** – (-$1200) we get paid, thanks to Solar Renewable Energy Credits.

HELIOS

In my architectural work, I endeavor to express both the mundane purpose and the essential meaning of a building in the design. So, when I became aware of solar building design, I immediately saw how the essence of solar energy could be expressed in the form and arrangement of buildings. This awareness happened swiftly in the 1970's, not only for me but for many architects and building designers. Our catalyst was the oil embargo and energy crisis

Figure 102 - Our completed home, HELIOS, with all electric Chevy Bolt in the background.

beginning in 1973--the year I opened my own architectural office in New Jersey. Architects in the southwest USA began experimenting with passive solar design, and I visited several of these homes in New Mexico. These solar designs inspired me, and I had the opportunity to create passive solar residential designs for clients in New Jersey in the late '70's and early '80's when there was a federal tax credit to encourage energy-efficient building design. Working with a mechanical engineer, we designed solar homes to capture the sun's energy for space heating: designs utilizing direct gain, sunrooms, mass walls, and rock beds for storage. I discovered that sloping glass on a sunspace is very difficult to shade in the summer months, and rock storage beds with fans are difficult to maintain.

Architects in those times felt like they were in the forefront of saving our country's energy future. One of my early clients was a solar pioneer who started a magazine called "Solar Age" to spread the gospel. But this explosion of interest in solar design was short-lived because the Reagan administration ended the tax credits and removed solar collectors from the White House. As a nation, we then went into a deep energy sleep – forgetting that we hadn't solved our energy crisis--and we became even more dependent on fossil fuels.

But some building designers (including me) didn't fall into that trap and instead continued to find ways to create solar buildings. My wife and I needed a new home for our young family at that time, so we searched the local area for a building site with a good view and south-facing slope, so we could plan a passive-solar, earth-sheltered home. Sussex County, New Jersey was quite rural in the early 1980's, with many active farms, so we quickly found the right site at the right price for our solar home.

Figure 103 - Model of a passive solar home design to help a client visualize.

Figure 104 - View from east side, showing the ample south-facing glass and overhangs for summer shade.

After several designs, I prepared the final drawings for an angular home with all rooms facing south, a two-story sunspace, and partially earth-sheltered on the north side. My engineer calculated that the passive solar design would save 65% of the energy needed to heat a similar size conventional home in the same location.

I acted as the general contractor and began construction in 1983. By the fall of 1984 (during a deep economic recession) our home was mostly complete, and we moved in with our first mortgage at an interest rate of 13 %. We named our home "Helios" (after the Greek Sun-God), and we never regretted building it. Helios has exceeded our expectations and we

Figure 105 - Our home viewed from the front lawn in early spring.

have continued to nurture it for the past 33 years so that it now has become a Zero Net Energy home that we think is as comfortable and aesthetically pleasing on the inside as it is functional.

Here is an inside view of the two-story sunspace on the south side. This area adds several hundred square feet of comfortable space for work and entertainment most of the year, while silently adding warmth and light into the home. The darker tile floor absorbs the sun's heat during the day to slowly release the warmth back at night. The trellis in front (not seen here) helps to keep the area cool in the summer by shading the high summer sun but allows the sun in during the winter.

Figure 106 - The sunroom adds heat and light into the home.

CARBON FOOTPRINT

I'm always amazed at how well the passive solar space heating at Helios works on sunny, cold winter days. As I'm writing this article, there is no heat on and it's 70 deg. F. inside while it's 32 deg. F. outside. Solar energy is absorbed by the concrete mass floor and exposed concrete walls which radiate it back to the home to keep the indoor temperature steady. But after sunset we need to add supplementary heat – in 1984 with a wood stove and later with a propane heater. So, our home was energy efficient but still needed wood or

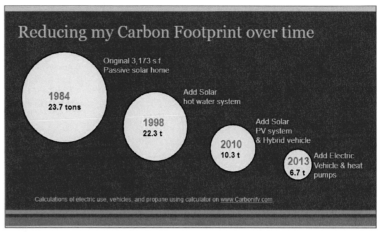

Figure 107 - For us, reducing carbon has been a journey.

fossil fuel or electricity from the grid. In 1984 our carbon footprint was 24 tons of CO_2 which was only slightly below the national average, and we hoped to reduce it in the future. (One ton of CO_2 inhabits a volume about the size of a 30-ft. spherical balloon).

In the 1990's, emerging scientific reports about the negative effect of carbon emissions jolted our awareness of how wood and fossil fuels adversely impact the environment, so Mary and I decided to try to lower our carbon footprint. In 1998, we added a solar hot water heating system on the main roof. This system has worked flawlessly to provide about 65% of our hot water needs over the year (less in winter but 100% in summer). This reduced our carbon footprint to 22 tons.

We tracked our monthly and yearly energy use and carbon footprint, but our footprint was still too high, so we decided to change the heating and cooling systems in our home to high efficiency mini-ductless heat pumps, which operate efficiently to minus 15 degrees F. With this decentralized system of heat pumps in each room, our home is now much more efficient, and we receive very inexpensive cooling and dehumidification as a bonus. As we became aware of the dire effects of carbon pollution on the global climate, we decided to offset our carbon footprint by purchasing "Carbon Offset Certificates" from non-profit organizations which promote global carbon reduction. At the end of each year, I went online to Carbonfund.org and B-E-F.org and used their calculators to figure the carbon cost of my footprint, including all vehicle and air travel miles. I then donate to these organizations to salve my conscience and help heal our planet.

Figure 108 - We purchase carbon credits to bring us to 0.

When federal tax credits plus state renewable energy credits became available in New Jersey, we installed a 5 kW solar PV system in 2008 to help power our home. Since the single sloping roofline of our 1984 home was designed for passive solar gain, not for solar collectors, we located the 22 collectors on the flat roof over the sunspace and the top of the south-facing ridge. By 2010 we reduced our carbon footprint to 10 tons, and by 2013, after installing two more high efficiency heat pumps, we further reduced our footprint to 6.7 tons.

After monitoring our energy use and solar production, we discovered Helios still required about 4 kW of power from the grid. So, in 2015 we added another 4 kW solar PV system, this time with micro inverters. After monitoring Helios for a full year, I confirmed that we had achieved a net zero energy home, plus we produced enough solar energy to power our electric vehicle.

The graph here shows the significant benefit of being grid tied. March through October we produce more power than we use, exporting the excess to the grid, building credits with the power company. Then November through February we can use those credits when heating loads are higher and solar production is lower.

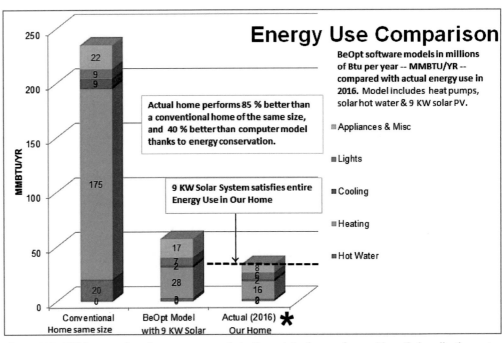

Figure 109 - Efficiency reduced our power needs to the point where solar could easily handle the rest.

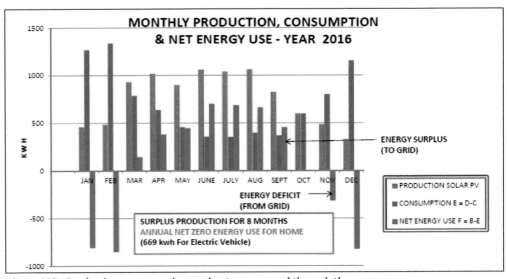

Figure 110 - Production, consumption, and net power used through the year.

ELECTRIC VEHICLES

As our distaste for fossil fuels increased, we decide to try an electric vehicle in 2012. We leased a Nissan LEAF (with 90 miles range) and installed a 240-volt charger in our garage. The LEAF is a very smooth sedan and extremely quiet. Our other car is a Prius with 50 mpg, so we used that for long-range driving and used the LEAF for all our local driving – which amounts to 60 % of our mileage.

What a liberating feeling to plug in our car in our garage using solar energy from our rooftop systems! This was true energy independence for us.

After our lease expired, we were ready for the new 2017 Chevy Bolt EV with 238 miles of range. It was my first GM car, so I wasn't sure what to expect, but the reviews were excellent. I'm happy to report that the Bolt exceeded my expectations – it is a blast to drive; very responsive at any speed; great handling; roomy interior; very quiet; and versatile for luggage and larger items – probably

Figure 111 - Our electric Chevy Bolt with reliable 240 miles of all electric range.

the best overall vehicle I've ever driven, and the best for reducing our carbon footprint and ensuring a cleaner world for our children and grandchildren.

With the Bolt's long range, we can drive a round trip to visit our daughter and family in Brooklyn without charging. And when we travel to Philadelphia to visit our son and family, we stop on the way for a 30-minute lunch at a mall which has a DC Fast charger, so we have plenty of range for the return trip.

To test the long-range ability of our electric vehicle, we bravely traveled to Montreal, Canada from New Jersey--a total round trip of 720 miles. Like Lewis & Clark setting off into uncharted territory, we hoped that the charging stations we planned to use would be available along the route. Pioneers need to take some risks so that those who follow in the future will have an easier journey.

With high speed highway driving and steep hills, we lost driving range faster than expected, so we needed to recharge three times in each direction. We planned the trip based on" Plug Share" charging station locations, and since there were no fast chargers on the last half of the northern journey, we had to rely on a slow Level 2 charger at a hotel on the way to Montreal.

Here are lessons learned from our long-range EV trip:

1. Plan your trip to only use fast Level 3 chargers if possible (Level 2 chargers are fine if you plan to stay at a hotel overnight).

2. Have a back-up plan for alternate charger locations in case one doesn't work or there is road work or bad weather causing re-routing.

3. High speed highway driving combined with uphill driving lowers range by about 17 %.

4. DC Fast chargers reliably provide 120 miles per hour of charging. Level 2 chargers reliably provide 25 miles per hour.

5. Expect your trip to take longer than normal due to charging stops – so plan meals or sightseeing at your charging locations and make that part of the experience.

SOLAR TOURS

As Helios evolved from a passive solar to an active solar home, I joined the American Solar Energy Society (ASES) and decided to participate in their annual National Solar Tour. Working with another solar enthusiast in my neighborhood, we hosted our first solar tour in autumn 2008. The event attracted a large, enthusiastic crowd of people from up to 60 miles away, and we've been hosting the national tours almost every year since. Visitors are

Figure 112 - Giving a tour of our home and fossil-free living with a college class.

especially interested in the solar PV systems and our electric vehicle.

To share my solar energy experiences with the community, I have opened Helios to tours by local college and high school groups, and I've lectured at our community college during their annual Earth Day celebration. It's wonderful to see the enthusiasm of young students and how well they understand the need for renewable energy and curtailing our carbon footprints.

To provide a central location to gather and share solar information, I founded the "Helios Energy Institute" with a board of directors composed of solar experts and mechanical engineers. We provide consulting services, and we're currently working on implementing a Community Solar project in our area of New Jersey. Additionally, we maintain a web page to share information and our story.

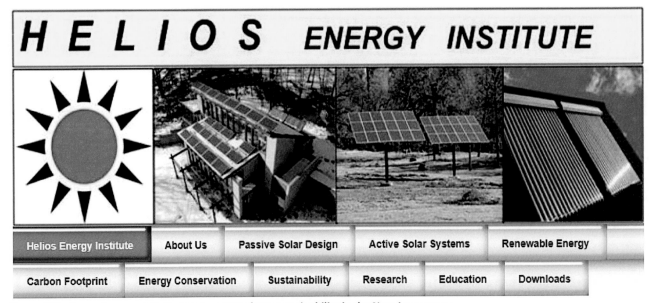

Figure 113 - Our web page at HeliosEnergy.org to share sustainability in the New Jersey area.

TECHNICAL DETAILS

HELIOS evolved over time from a passive solar home, to a home which also reduced its hot water energy needs, to a home with reduced heating and cooling energy loads, and finally to a net zero energy home. Helios was first constructed in 1984 as a passive solar home with the intent of employing proper building orientation, partial earth sheltering,

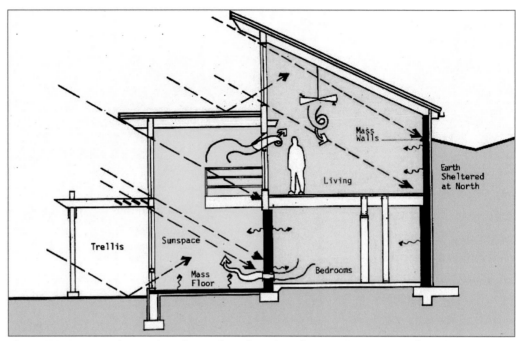

Figure 114- Cut away view to show the passive solar gain and storage in concrete walls and floors.

solar gain through a two-story sunspace and south-facing windows, and building mass to reduce the heating load in rural northwestern New Jersey (5,397 heating degree days). This design resulted in a 2,751 sq. ft. home and 420 sq. ft. sunspace, which consumed only 35% of the energy needed for heating a comparable-size conventional home. In 1998, a solar hot water system was added to provide 65% of the hot water energy needs. With the advent of high-efficiency mini-ductless heat pumps, this equipment was installed in several rooms to replace the existing electric baseboard and propane heaters in the home. When solar photovoltaic (PV) systems became more affordable in 2008, a 4.7 kW (DC) system was installed and, with further economies in solar PV, an additional 4.1 kW (DC) system was installed in 2015. The result is a net-zero energy home which has been carefully monitored over a full year of actual use to verify that all energy for space heating, cooling, hot water, and plug loads is provided by solar energy.

When off-site "Community Solar" systems are permitted in New Jersey, we plan to purchase the necessary energy from a local community solar system to fully satisfy the energy needs of our electric vehicle (estimated to be 3,000 kWh/year).

Materials & Design Strategy

Mass walls and floors are exposed poured concrete (painted a dark color). Interior walls are mostly cedar and redwood (with minimum use of gypsum board on walls). Ceilings on all floors are exposed spruce-pine-fir tongue and groove wood decking. Exterior masonry and wood stud walls are covered with EIFS acrylic stucco finish. Landscaping is natural native plant gardens, existing oak and maple trees, with a small lawn area on the south side.

The two-story sunspace is unheated and acts as a buffer to the conditioned spaces behind. The north side of the home is earth sheltered (8" thick reinforced concrete wall with 2" rigid insulation on exterior) and the interior of the exposed concrete wall acts as a mass wall to store solar heat during a sunny winter day. Deep roof overhangs protect from overheating in the summer months. The solar heating performance was carefully monitored and calculated to be 65% better than a similar-size conventional home in this location.

Figure 115 - The conditioned space gains light and heat from the sun space and upper windows.

Natural convection air flows are induced by opening upper windows at the third-floor loft area in the winter and opening low windows at the sunspace in the summer. Conventional exhaust fans are in the kitchen and bathrooms. In the future, a heat-recovery ventilation system will be installed.

Solar Energy Systems

The first solar system (installed in 2008) comprises 22 REC Model REC215 solar collectors with an SMA Sunny Boy 5000US inverter which provides 4.7 kW (nominal power). The newer 2015 solar system includes 16 REC model REC255PE solar collectors with SMA micro inverters and provides 4.1 kW (nominal power). Total system comprises 8.8 kW (DC) nominal power. In addition, there is a solar hot water system which includes two 4'x8' flat-plate collectors and an 80-gal. hot water storage tank which produces about 2,080 kWh/year. Annual solar energy generated = 9,196 kWh.

Figure 116 - Aerial drone image showing the solar panels mounted on the roof.

Energy Storage

Hot water storage carries over for several days during the spring and summer but is quickly dissipated in fall and winter (based on solar hot water monitor readings). When there is full sun, passive solar heating in winter months is stored in thermal mass walls and floors. Based on 33 years of observation, on a cold January day (20 deg. F. outside) with full sun, supplementary space heating in the living rooms is not needed until about 6pm. Solar PV is grid-tied and there is no battery storage. The New Jersey SREC program (Solar Renewable Energy Credits) currently provides payment of about $200. per each 1000 kWh of solar production, and the electricity is "net metered," so our utility credits us for all solar production which is sent back to the grid. In the future we may consider adding a Tesla Power Wall type battery if the price drops. It can be used for charging our electric car during evening hours and be available for emergency back-up during a power outage. Our electric vehicle could also be a source of emergency power in the future, and I'd be interested in a reliable inverter (not yet available in the US) to connect my vehicle to our electric panel.

Walls & Roof

Insulation and windows were provided per 1984 building standards. These significantly underperform compared to current standards. Roof insulation is rigid 4.5" thick nailbase polyiso insulation over exposed wood decking. Frame walls are 1-1/2" rigid insulation & EIFS system over 2x6 studs @ 16" on center with 5-1/2" fiberglass blanket insulation. Rear exterior wall is 2" XPS insulation over 8 to 12" thick poured concrete retaining wall, forming the earth-sheltered north wall of our home. The south facade has a 30' wide x 2 story high sunspace, which buffers the home from outside winter weather conditions. Three-foot roof overhangs provide shading for summer sun but allow full sun in winter for solar gain.

Mechanical Systems

We use three high efficiency Fujitsu Halcyon mini-ductless-split-system heat pumps for heating & cooling. The units are located to serve the living room, kitchen, and master bedroom. Domestic hot water is from a GE Geospring Heat Pump. This is a backup to inter-connected solar hot water heating. For lighting, we are moving to LED lighting.

Software Tools

Heating performance was monitored for several years, and energy modeling substantiated a 65% benefit over a conventional home in the same location.

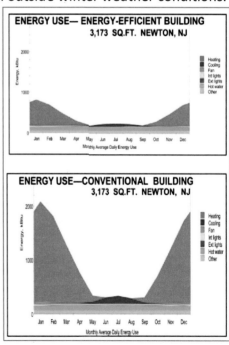

Figure 117 - Energy 10 software modeling.

Originally, "Energy 10" software was used to model the original passive solar design with conventional heating and cooling systems. Later "BeOpt" software (Building Energy Optimization - with "EnergyPlus" as the simulation engine) was used to model the home retrofitted with heat pumps and solar PV.

LESSONS LEARNED

I'm pleased to share these important lessons learned during the evolution of our home:

1. Earth sheltering is very helpful to reduce heat loss and infiltration (but humidity needs to be controlled).

2. A two-story sunspace with vertical glass is an excellent buffer for conditioned spaces of the home.

Figure 118 - Overhead view of our home, HELIOS; with Chevy Bolt & garage with studio above.

3. Properly sized concrete mass floors and mass walls work well to stabilize indoor temperatures in winter and summer--the more mass the better.

4. A solar hot water system with a heat pump water heater as backup works very well to provide all hot water needs.

5. A solar PV system (with micro inverters in locations with some trees) can provide **all** the power needed to achieve a net zero energy home.

6. Decentralized heating and cooling utilizing high-efficiency, mini-ductless heat pumps is an excellent way to achieve high efficiency.

7. A solar PV system should be over-sized to provide power for future electric vehicles.

Living in a net zero energy home is very empowering! My wife and I live fossil fuel free, and we're energy independent. We know that we're not contributing to carbon pollution and are doing our part to leave our grandchildren a clean and healthy planet.

The passive solar components of our home allow us to live in harmony with nature's rhythms (low sun angle in winter, and high angle in summer). And the active solar systems allow us to precisely monitor our home's energy performance and improve energy conservation.

For everyone interested in employing renewable energy for a healthy planet, we learned we can reduce energy consumption in our homes and businesses, and we can do something, however large or small, to reduce our use of fossil fuels. Then we can offset the rest of our carbon footprints by purchasing carbon offset certificates. My wife and I are happy to be among the pioneers in this work, and the future looks bright and clean!

Alan Spector is a registered architect in New Jersey, and a member of the American Institute of Architects. He received his Master's degree in Architecture from Harvard University, and his Bachelor's degree in Architecture from Columbia University. Alan and his wife Mary live in their net zero energy home in Northern New Jersey.

Figure 120 - Informational graphic of our home and energy systems.

Figure 119 - Additional information on our home and its energy systems.

For additional information see:

	Zero Net Energy Solar Home – Helios	https://www.youtube.com/watch?v=0TaPndpKIOo&t=8s
or	Solar Pioneers - Alan Spector	https://www.youtube.com/watch?v=15YsVyAGMTU

Chapter 8 - Ten Years with a Bucket of Shit: An Attempt to Live Consistent with My Values.

by Pete Schwartz

Sept. 2, 2017. Yesterday, I installed a water toilet, the kind that mixes purified water with feces and urine to be returned to a processing plant, thus ending my 10 years living with a dry bucket toilet that diverted urine directly to our plants with the rest of the grey water. My family's poop was composted in the back yard with dead vegetation and then spread on fruit trees. My friends and students asked me why I replaced my bucket toilet. Maybe it was because my super partner, my daughter who constructed it with me at age four, is now an embarrassed teenager who won't bring her friends home any more. Or because my extended family is disgusted at how we "live in our filth." Or simply that after about 250 hours of shoveling shit, maybe I can find something more creative to do with my time. But maybe I was motivated by one student's final exam essay describing their transformation through the appropriate technology class I teach at California Polytechnic State University (Cal Poly). He said that he never wanted to stop changing and growing. He recounted the class trip to my house where I indicated that I may have learned what I needed to learn from the bucket toilet, and it was time to move on. My willingness to change identified my commitment to innovate more than my bucket toilet did. Besides, I don't use the water toilet. I use the other bucket toilet in the outdoor shower.

Quick Facts:
>**Net Energy Emissions** – 6600 lbs./year – 82% less than the typical American family
>**Home** – 1920's kit home, 700 sq. ft. in Coastal CA, USA. Another family lives in a 440-sq.ft. annex
>**Technologies Employed** – Passive solar, solar hot water, home gardening, bicycling, grey water
>**Annual Energy Costs** – $775

But isn't the more pertinent question: what motivated me to shit in a bucket in my house in the first place? It wasn't actually my idea. That idea came from my daughter's mother. She was so excited about getting a bucket toilet that she researched and mailed away for one. I read The Humanure Handbook by Joseph C. Jenkins and figured out how to make it work. And then I built one for inside the house... after we split up.

But *really*, how did this whole thing happen? Previously a lecturer at Cal Poly, I was hired as permanent faculty in 2003, so I bought my first house. I could never have guessed what an effect that would have on me or on my lifestyle. I'd spent most of my 40 years bitching about the way people irresponsibly use our land and resources. I was struck by the realization that I needed to hold myself accountable for the choices I would make as a homeowner. So, I began making choices that felt consistent with my previous rants; choices that in the last decade changed many things in the house and in me. I started with the toilet and things developed from there.

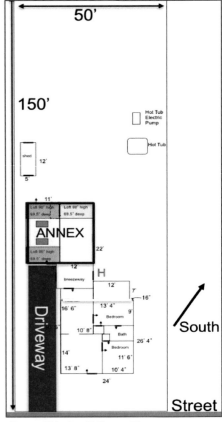

Figure 121 - Lot & House Plot

Composting

The toilet is made from a 15-gallon bucket under a wooden platform. An air tight seal is made between the bucket and the platform as well as between the platform and the lid. For this reason, the lid rests directly on the platform rather than on the seat. I made a urine diverter from a funnel, allowing the urine to join the grey water and be deposited directly onto the plants. We quickly found that urine crystals would accumulate and block the urine diverter drain. We remedied this by flushing the urine diverter (before and after urinating) with about 10 oz. of water via a flush tube connected to a plastic peanut butter jar. After using the toilet, one would also cover their poop with leaves to reduce smell and moisture.

Figure 122 - Bucket Toilet System

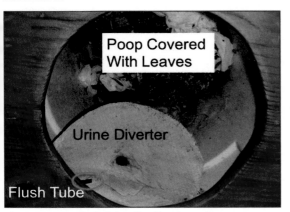

Figure 123 - Technical details of bucket toilet.

I learned to compost the poop in the back yard with layers of yard waste and compost from the kitchen. Every two weeks, I built a 4' high compost pile onto the previous pile, following a row that would repeat itself in about a year. With proper layering and ratio of poop to dry plant matter to water, the thermophilic bacteria would drive the temperature to 140° F for a few days resulting in considerable condensate and a reduction to half its original size. Rather than turning the pile over to make sure that all the poop was exposed to this temperature, I only put the poop near the middle of the pile. I could write a book about this topic alone, but I don't have to. Please do read the <u>Humanure Handbook</u>. Even if you don't want a bucket toilet, it's insightful and funny.

Figure 124 - Composting yard, kitchen, and human wastes. Sheet (not necessary) but shows condensate.

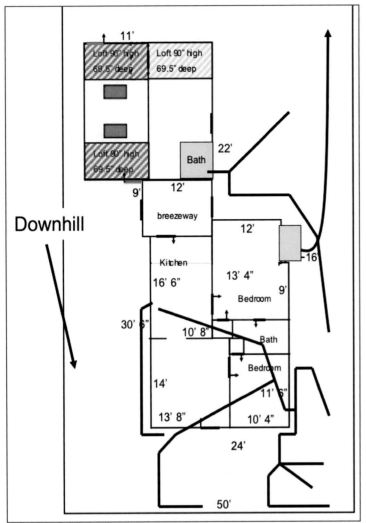

Figure 125 - House plan with grey water system.

Grey Water System

The toilet installation led to installing a grey water system to maximally use our water and water-born nutrients. I installed a branched grey water system. I was able to run 1 ½" ABS piping at a standard ¼" per foot minimum slope into a branching network, terminating in many buried plastic receptacles that I covered with pieces of brick for (relatively) easy removal.

Figure 126 - Brick covering grey water receptacle.

This system allows us to conserve water by using waste water to irrigate our plants and grow some of our food. The virtue of a branched grey water system is that there are no moving parts, filters, or process requirements. The down side is that you don't precisely control the water flow, and wet earth can seep into a receptacle, requiring it to be dug up and emptied.

The system works quite well and contributes to our goal of reducing our impact on the environment. From here we went on to design and install several improvements to our living space the bring us closer to net zero.

One sees one pipe heading uphill from the blue shaded laundry area in Figure 125. This is possible because the washing machine has a pump.

Washing Machine Discharge

Figure 127 - Washer drain, brief uphill to grey water.

The House and Passive Solar

Figure 128 - Evolution of my home over the years, from kit house, to when we moved in to present.

My house, built in 1924, likely from a Sears Roebuck kit, is about 670 sq. ft. with a 440 sq. ft. annex--where another family lives--on a long, thin 7500 sq. ft. lot. The back yard of the house faces mostly south. In an ideal design, south-facing windows let in the winter sun and provide some heat in winter. However, the house had no windows facing south. I had large windows put in the south face of the annex, a sliding glass door put in the bedroom, and a light pipe put in the kitchen. We extended the peak of the roof southward, allowing a row of windows to be placed at the top and finished the attic into a loft for my daughter. This loft roof extension has significant overhang to shade the summer sun, while allowing penetration of the low-angle winter sun. We also removed a good piece of attic flooring, allowing sunlight to come into the northern-most portion of the house. Potentially, the greatest difference was the way we used the master bedroom. It was initially cold, dark, and dead. The sliding glass door we installed made it a bright, warm passage to the back yard, transforming it to the center of our living space. We put two queen-sized mattresses on the floor under

Figure 129 - Adding considerable glass on south side for solar gain & light. In the summer (bottom), deciduous fruit trees shade the house.

a trapeze, and it became the gym, library, and entertainment center. We added decidious trees to the south to shade and cool the house in the summer. The transformation inside the home was far greater than what can be seen on the outside. The loft and opened living area are shown next.

Figure 131 - Loft area serves as daughter's room & allows southern light to penetrate.   Figure 132 - Living room bright, with natural light.

The extra light and brightness in the master bedroom transformed it into a favorite gathering place for my kids and their friends.

San Luis Obispo has a mild climate, with only a few frosts each year, so it is possible to get by without a heating system. There are additional passive solar strategies that make the home more comfortable. The corner facing South (labeled "H" for "hot" in Figure 121) reaches 120° F even on cold winter days, as the white walls and low-e glass function as a solar concentrator. Additionally, the south face of the annex was unpleasantly hot in the summer. I planted a fig tree south of the annex and a peach tree in the "H" corner, and grapes for the trellis. These trees provide shade (and fruit) in the summer, and strong sunlight in the winter, when the heat is appreciated.

Figure 130 - Master bedroom now bright and inviting with double glass doors.

I put a polycarbonate enclosure in the area that receives the most sun and pump the warm air into the kitchen on cold sunny days.

Figure 133 - Passive solar features added to house.

I also improved the thermal qualities of the house by more traditional means. I had insulation blown into the void between the external skin and the interior plaster, and I had an energy audit. We sealed the lights and electrical outlets and insulated the ceiling. When it came time to reroof the house, I chose white shingles to reduce heat loss in the winter, and solar gain in the summer. I also had polycarbonate storm windows installed on the inside of my windows.

Figure 134 - Pear tree shortly after planting.

The southwest side of the house gets beaten on by the evening sun in the summer. I planted pear trees at the base of the house to climb this wall. They reached the roof in three to four years, shading this wall significantly. Now, both the fig and the pear trees are fixed to the roof eves with large eye-bolts, and will continue to grow onto the roof, providing a green roof with the roots in the ground.

One mistake I made was to plant the trees too close to the house, which resulted in cracks in my foundation a few years later. I had to dig down beneath the bottom of the foundation, cutting the roots that went under the house. Then I laid another layer of concrete with rebar. Finally, I planted another row of fruit trees considerably further from the house. When the new trees grow big enough, I may remove the older trees closer to the house.

Figure 135- After several years the pear tree has grown.

Initially, the property had mostly ornamental plants. About 7 years ago, I decided that if I support plants, they should support us as well. I replaced almost all the ornamentals with food-bearing plants. We don't grow too many vegetables because we have a wonderful weekly farmer's market in town. But I planted fruit trees and grafted onto some of the existing trees. For instance, our ornamental plum, now produces plums and pluots.

Figure 136 - Grafted plums and pluots for eating.

Figure 137 - Grapes & Figs provide shade and food.

We grow lots of passion fruit, guavas, figs, grapes, pears, and all kinds of stone fruit. We make several gallons of jam a year – mostly from fruit that was already partially eaten by animals – opossums, raccoons, rats, and birds. All these animals were already there before I started growing fruit. Likely, they live better now.

Figure 138 - Some of the passion fruit we grew and then turned into jam.

Solar Shower

The guys working on the house and I designed and built a solar-heated outdoor shower. The picture below was taken before I replaced the trumpet vine with guavas, strawberry guavas, pomegranate, and thornless blackberries (watered with shower runoff) that one sees in previous pictures of the back yard. Two solar thermal panels are made from copper tubing soldered to a thin copper sheet, painted black and sealed under glass. A small photovoltaic panel powers an electric pump that circulates the water through a discarded water heater when the sun is up. The water is still hot in the morning. A discarded concrete wash basin served as a tub for my daughter, who preferred bathing over showering.

Figure 139 - Our solar shower with water heated via solar thermal panels and my daughter's tub we removed from the house.

Showering is so much better outside, under the stars. There are two shower heads and enough room for the whole family, although it's usually just my partner and me. After the water's off, we're quickly dry because there's no cloud of water vapor hanging over us. Even in January it's amazing to run out into 35 °F to get blasted with scalding water. The colder the air, the hotter the water one can bear; so, it's like being steam cleaned. On cloudy winter days, the showering is cold and fast… and lonely.

In 2006 my daughter's mother noted that I was more excited about designing the solar shower in the back yard (the thermal and mechanical calculations) than about my nanotechnology research in the laboratory. She suggested I think about that. I was planning my coming sabbatical year with a nanotechnology group at New York University. I considered her suggestion for about 20 minutes and chose to spend the next year at UC Berkeley studying strategies for global sustainability.

I struggled with my feeling of hypocrisy in owning a hot tub and sought ways to heat it without burning natural gas. I bought a transparent pool cover--like heavy-duty bubble wrap--for it. This brought the water temperature to close to 100 °F. However, I had started using ozone instead of chlorine to purify the water, and in a week, enough algae grew to turn the tub, cover, and water black. After a few more failed efforts, I threw a pail of tadpoles in the water, solving my personal conflict about owning the tub. Then there were mosquitoes. I got guppies. They died when the water got cold. I facilitate Cal Poly's Student Experimental Farm, where the most vibrant project is PolyPonics, the Aquaponics Club. I took some minnows from them, which multiplied to be an estimated population of 1000 in about a year. Then I brought home some tilapia and catfish from campus. Then I got three bass. I couldn't get the bass to eat fish food until a month later, when all the minnows were gone.

Aquaponics and the Lawn

I find aquaponics to be a great idea. However, it typically requires pumps, pipes, tanks, and power sources that make it inappropriate for the world's poor. Thus, we experiment at home with growing plants in the stagnant (but aeriated) water of the hot tub. We've managed to grow a forest of water lettuce, but you can't eat it. I know; I've tried. We've been unsuccessful with peppers, tomatoes, and passion fruit. I'm presently trying traditional pond plants including water cress, water hyacinth, and lotus.

Figure 140 - Our first attempts at aquaponics, it's been *fun,* and we are learning, but we have yet to provide food.

Grass is likely the largest irrigated crop in the US. Manicured lawns consume water, petrochemical fertilizers, and lots of effort. Additionally, many people put their grass clippings in the landfill, which produces methane, a potent greenhouse gas. The lawn was the first thing that had to go, but what are the options? The first option we tried was to do nothing, letting Nature do it by herself. This worked out great but resulted in tall grass that dried to be a fire hazard, lovely wild plants that spread seeds angering our neighbor, and nasty foxtails and bur clover. I introduced and neglected several ground covers to see which variety prevailed. Dymondia was one that proved to be very hardy. However, in a bad drought it dies back. The overall winner was a chunk of Cape Weed (Arctotheca Calendula) that I took growing from a neighbor's yard, growing into the road. It dies back a little during the summer but comes back when it rains. It now covers most of my yard. In researching the plant for this chapter, I discovered that Cape Weed has recently been classified as California invasive!... bummer. However, I'm quite sure that the variety I have does not go to seed, so its invasive speed is likely only about one meter per year rather than the hundreds of miles per year if it were to have fertile seeds. Shown from left to right is what naturally grew up, dymondia, drought cape weed, and lush cape weed both dry and wet.

Figure 141 - Natural ground cover, dymondia (at our wedding), dry cape weed, and lush cape weed.

Retained Heat Cooking

Another important professional/personal crossover is our retained heat cooker (RHC, or "hot box", or "Magic Basket") that I first saw at Tryon Community Farm, an intentional community in Portland. If you need to stew something like beans, you can bring them to a boil and then put them in an insulated chamber to continue cooking without added heat. We have successfully done this both inside the oven and inside a cooler. We routinely cook beans this way in large amounts to freeze. It was not much of a stretch to put an electric heater in the pot directly connected to a 100 W photovoltaic solar panel to develop (what I think) is an improved solar cooker for the global poor. My students and I recently published this work in Development Engineering:
http://www.sciencedirect.com/science/article/pii/S2352728516300653

Figure 142 - Retained heat cooking in an oven and cooler; with added PV, a possible cooking solution for the global poor.

I don't pretend to think that my practices at home are the answer to society's problems. However, I think that the answer to our challenges lies in the collective discussion that takes place when humanity innovates. When I bring students to my home, I don't suggest they should compost their family's feces in the back yard or take on any of my individual practices. My message is only that when you see how others do something, you might consider, "What if I did it differently?" and walk down that path as far as you consider beneficial. For instance, when a friend asked if she could live in the back yard in a teepee, I might have been inclined to ask why. However, my response was, "Why not?" I think that this practice of versatility itself may be the most beneficial.

Figure 143- Friend in a teepee

Some Numbers

My students and I calculate our carbon footprints for my energy and appropriate technology classes. One can use any number of websites (such as this one at UC Berkeley: http://coolclimate.berkeley.edu/carboncalculator). The four of us together emit about 16 tons of CO_2 per year, making us about the same as the average world dweller, or ¼ that of the average US citizen of about 16 tons of CO_2 per person per year. Note this figure includes food and shopping, which is largely omitted from the other chapters.

Automobiles: We bicycle exclusively around town, thus providing transportation, exercise, and sometimes more adventure than we would like. We drive our 1996 Subaru Outback to the beach and other out-of-town trips logging about 4000 miles per year, corresponding to 140 gallons of gas, or about 1.4 tons of emitted CO_2. We fly to Phoenix every year to visit family over winter break.

Electricity: We use very little electricity, about 100 kWh per month, which costs about $14/month. As the marginal electricity in California is generated with natural gas, it may be reasonable to assign about 1/3 kg of CO_2 per kWh, which means we emit about 400 kg of CO_2 per year from electricity. My students and I installed about 1600 W of solar panels last year. It's not grid connected, but rather for experimenting. The panels heat the hot tub in the winter so the tilapia don't die, and we also have an electric immersion heater in our natural gas water tank and may use this capacity to trickle charge an electric car someday.

Natural gas: My family uses about $10 worth of natural gas per month, about 30 therms. Per year, this is about 360 therms or about 36,000 MJ. About 25 grams of CO_2 per MJ equates to 900 kg of CO_2 per year for cooking and heating water. Additionally, we have a small hydronic heater that we use from time to time when it's very cold…well, "California cold," that is. (I grew up in Buffalo NY). A pump circulates water from the hot water tank through a hydronic baseboard radiator.

Embodied energy/carbon: The embodied energy and emissions in putting up a building is equivalent to about 15 years of operational emissions for a regular building and more for an energy efficient building. The energy it takes to manufacture an internal combustion automobile is about equal to that of 15,000 miles of driving. Electric cars require more because of the batteries. And the recent surge in lithium mining has dramatically increased the cost of lithium, and likely also the environmental cost, because the remaining lithium is more difficult to mine. How will we account for all this? It's complicated. However, it is simple to understand that everything we use, buy, dispose of has an environmental cost. One of the nice things about my house is that it was already here. It's been 15 years, and I'll likely get another 30 years of use out of it before I'm … "done," what will happen then? It's nice to think that I have transformed the house into something that people will value into the next century. However, this may be wishful thinking. In a place where property goes for about $100 per square foot, or over $4 million/acre, most of the value of my home is in the plot and location. It is likely that future buyers would optimize their investment by building a much larger house. Thus, what I have here may simply be a reasonable experiment. However, it still provides talking points, stimulating the conversation to innovate society.

Conflict: There's very little written about military-related emissions, and I support these emissions by paying taxes and educating engineers. If we are to build a sustainable future, we must consider what the global costs are of maintaining the US military-industrial complex. The budget of the defense department is presently only 5% of the GDP, but has been much higher at other times, and the military uses a greater portion of its budget for petroleum than the rest country. Additionally, much of the rationale for enlarging our economy and the related emissions is to maintain global dominance. Certainly, there is a significant global environmental cost to not getting along. This extends to all levels of society--from world wars to my daughter's complicated transportation schedule--which resulted because I wasn't able make things work with her mother. I'm embarrassed to admit that I tried to prevent my neighbors from putting in an apartment complex in 2008 and only succeeded in making enemies for life, never mind that infill, using unbuilt space, is a legitimate sustainability strategy. Conflicts cost us dearly: financially, environmentally, and emotionally. So, what do we do? We remember to listen and empathize. We make empathy and collaboration part of the discussion – for me, part of the classes I teach. I think I'm getting better, but change is hard.

Food: We eat as much as the average person, so this is the largest part of our energy/carbon footprint. We grow about half of our fruit. We also have a "vegan-leaning" diet. We try to eat for a healthy body and healthy planet. We don't make rules, but rather recognize cost – just as someone might enjoy a very expensive meal once in a great while, we might sparingly enjoy a meal with a larger environmental impact. I liken it to driving a car: It's not whether you own one or not, but how much you emit. Cars only emit when you drive them, just as the environmental impact of eating beef scales with the amount of beef you eat. I was a vegetarian for more than 30 years. When I started teaching and studying sustainability, I was disappointed to learn that milk products have as high an environmental footprint as beef, where the environmental footprint of poultry is much lower. Consequently, we eat some poultry and milk products, but not so much.

Buying stuff (we don't need):
Much of our consumer society is dedicated to digging things up to be buried shortly thereafter in landfills. I don't pretend to not be a consumer, but we make this awareness part of our family discussion. Part of the holiday tradition used to be watching videos of Black Friday fist fights at Walmart and talking about what

Figure 144 - The sofa trip from curb, across town, to my daughter's loft.

was behind the commercials that constantly assault us and our children. In trying to strike a balance between giving in and tyrannically imposing my values on my children (which probably wouldn't work anyway), we discuss about where things come from and where they go. We keep our things until we can use them again, give them to someone who needs them, or drop them at the free box in town. Once, while I walked my daughter to her mother's house, she expressed an interest in a sofa on the sidewalk saying, "I know you don't like to be a consumer, but I'd like that in my loft." I returned home, and my partner and I brought the sofa home from across town--shown in the picture sequence in Figure 144.

Discussion with the kids used to be easier. My son is forever enamored with the latest robotic gadgetry, and my daughter insists that I should drive her wherever she wants, like all the other kids' parents do, because bicycling is so embarrassing... "Homeless people bicycle." At present, I've made a deal with her that I'll buy a used Nissan LEAF and charge it with the panels on our house if she'll ride her bicycle when it is reasonable. We differ on the meaning of the word "reasonable."

Conclusions:

There are many things we can do to benefit our planet. Some of them are technical, but to me much boils down to the life choices we make. The largest challenges related to our/my practices and interventions is in fielding others' questions about our lives. However, this may be where we find the greatest potential for positive impact. We do not know what lies ahead of us, technologically, socially, politically, environmentally. Which way will we go? This steering lies in our discussions. What do we think? How do we feel? What do we prioritize?

Imagine you are at a party: They're talking about the Lakers. You might just listen, but you could suggest it's a shame that all this good food and drink will become bodily waste to be mixed with purified drinking water and transported away. And you're curious if instead you might deposit the resources your body has processed outside somewhere so as to be of value to the family's ecosystem. Or maybe you'll bust out a 10 pack of LED lightbulbs and replace the incandescent ones while explaining the difference. Or maybe you'll find some other way to upcycle the conversation... Or you could talk about the Lakers.

Let me know how it goes.

Chapter 9 – RISE

by Rita Hennessy and Sean Palmer

"I have always supported nuclear energy so long as it is in the form of fusion and parked 8.3 minutes away (as the photon flies)! Yes, this is Sol, our sun, and the star that we circle once a year."   -Sean Palmer

Quick Facts:
> **Net Energy Emissions** – 3,400 lbs. or 92% less than the typical American family
> **Home** – New Passive solar home, 2,139 square feet in West Virginia Panhandle, USA
> **Technologies Employed** – Passive solar, solar PV, super insulation, electric vehicle, heat pump water heater, masonry heater and woodstove
> **Annual Energy Costs** – $425, down 81%

Background

I grew up in the 1960s, with thousands of acres of mountainous woodlands, spotted with ponds and streams as my backyard, including a state park and a municipal watershed. These surroundings played a pivotal role in my understanding of the relationships between plants, animals, the environment, and the Earth. I could see the impact that our species was having on the planet at the time. I started collecting aluminum cans and recycling them. Other materials came into play such as paper, glass, and other metals. These are finite resources, I thought, and recycling them just makes sense so that less of the planet must be disturbed.

One of my first jobs was with a solar cell manufacturer. I thought this was a great "new" concept. From there I worked as a carpenter, building custom homes in the 80s - large, energy inefficient "McMansions," which was not the best concept, but was indicative of the economy and the American mindset. My experience led me to dream about the possibility of building my own passive solar home.

In the early 90s, I met my future wife, Rita. We bought a small, 1940's, Cape Cod style house. It was a quaint starter home. There was no insulation in the walls--just plaster on brick. An oil furnace with hot water baseboard radiators and a woodstove provided heat in the winter. We fixed it up over the 20 years we lived there, gardened, composted, and recycled, while building a savings for the future. In 2006, to be

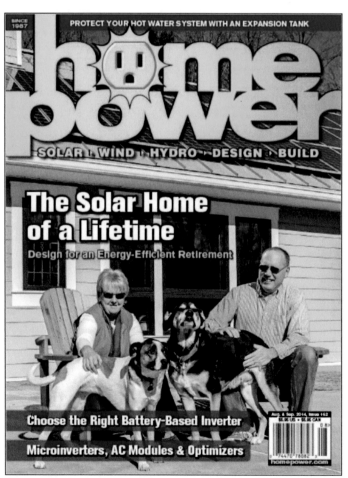

Figure 145 - Cover of Home Power magazine, Issue 162, Aug/Sep 2014; copyright by Matt Hovermale

even greener and start reducing our carbon footprint, we bought a Toyota Prius. In 11 years, I learned efficient driving techniques and averaged 60 mpg over the 300,000 miles that I drove.

Then, in 2008, we began to put together our energy efficient future plans. We defined what we wanted in a tract of land to include qualities such as good southern exposure for future solar usage, located near public park land, a blend of forest and open space for a managed wood lot and orchard, and the house site. We spent two years looking at various parcels before landing on the lot we purchased in December 2010.

Design

We began looking for architects with a specialty in passive solar design and found Debbie Colman, owner and architect of Sun Plans, Inc[19]. We spent the year working with her to develop the design. We aimed for a house that was primarily one story and about 2000 sq. ft. I focused on efficiencies, while Rita focused on making it aesthetically pleasing to the eye. Debbie connected us with an energy specialist to determine what size heating unit to install. Solar gain must be taken into consideration, and then we were told that it must be sized based as if the sun wasn't shining. Cloudy days do happen. This is where the base line efficiency of the house comes to light. How well is it insulated? How well is the envelope sealed? Based on the answers of these questions, we chose two-small ductless mini-split units with a small ducted unit--all tied to a small heat pump outside. This seems like an incredibly small heating load, for the 2,139 sq. ft. house that we designed.

Rather than SIPs, which can be expensive, our house is designed with a double 2X4 exterior wall. This produces a wall that is 9" thick from inside to outside and has greater than an R30 insulation value. The studs are staggered to eliminate thermal bridging between the walls. The windows are triple pane to offer the best balance between insulation and Solar Heat Gain Coefficient (SHGC). The SHGC is the measure of how much sunlight makes it through the glass. With a passive solar design, you want light to pass through to be absorbed and turned into heat. Such a design incorporates thermal mass to store the heat energy to radiate it back when the sun is not shining. We elected to use an insulated 4" concrete slab to accomplish this.

The ceilings were designed to be better than R60. The whole envelope was sprayed with closed-cell, expanding foam insulation. A ½" – 1" layer was used to seal the house against air leaks. This also has the benefit of strengthening the structure by bonding all the studs, joists, and rafters to the exterior sheathing and roof decking.

With access to our own wood supply, we installed a masonry wood heater. We purchased a kit for the core unit and added a veneer of fieldstone from our property, along with some bricks to finish it off.

Figure 146 - The Phoenix, includes ample masonry for heat storage.

Approximately 20 - 30 feet of flue are within the massive structure to capture the heat given off by the fire and the thermal mass of the unit radiates the heat for 24 hours after the fire burns out. The unit, a Phoenix, supplied by Empire Masonry Heaters, is a small masonry heater that burns very efficiently and cleanly. In many calculations, woodstoves are considered net neutral, as the carbon cycle is extremely short from a managed wood lot. This is carbon that is already in the carbon cycle above ground, as opposed to fossil fuels that have been stashed away deep underground for millions of years and then pulled out and added to the atmosphere by burning them. This causes CO_2 levels to rise, and the global temperature soon follows. Pellet stoves are not considered net neutral because there is a lot of energy required to manufacture the pelletized fuel.

On the other end of the spectrum, cooling during the summer months was considered. With the large expanse of southern windows to collect sunlight in the cool months, shading must be used during the summer to prevent overheating. This is designed into the house using large overhangs. These are sized so that at noon, on the winter solstice, the sun shines deeply into the house, while on the summer solstice, the entire south wall is in shade. It is magical to see the sun's path across the southern sky raise and lower during the year and how the house, by design, passively takes advantage of this phenomenon. The interior of the house is designed with a cathedral ceiling in the main living area with a clerestory at the top. By opening the windows high in the clerestory and the lower windows on the main floor at night, a thermo-siphon begins. The hot air that has risen to the top of the living area exits the clerestory and this pulls cool air in through the lower windows. This cools the slab and the rest of the house. In the morning, the windows are closed to keep the coolness in the house. On really hot days and times when

the nighttime low temperatures remain above comfortable levels, we pull the shades to keep out some of the light and heat during the day and turn on the air conditioners or dehumidifier of the mini-splits.

We were also able to use a large amount of reclaimed lumber for an exposed timber feature in the house. Lumber was reclaimed from a local factory that was built in 1835 and torn down in 2011.

Figure 147 - Interior of main living area with the sun penetrating deep in the house to warm the dark floors on January. Notice the reclaimed lumber above, and a mini-split hanging on the wall to the right.

Figure 148 - Solar noon in summer shows the shadow of the overhang, demonstrating passive cooling design.

One of the largest electrical demands in a house is making water hot. We considered solar hot water, but the ROI was not worth the investment for us. Instead, we chose a hybrid hot water heater. It is a combination of a heat pump and a conventional resistance hot water heater. Running in straight heat pump mode, it uses about a third of the energy used by a conventional unit. It is installed within the living envelope of the house. We learned that operating it in heat pump mode during the summer is not only efficient, but it pulls heat from the living space while dehumidifying the air. This was an unexpected benefit. During the colder months, we switch the hybrid hot water heater back to conventional mode so as not to rob the house of heat. We produce enough surplus electricity that it is not a concern for the extra power needed for resistance heating.

Solar Array

During construction in 2012, we installed 27 X 235-watt solar panels--a 6.345 kW system that is grid tied. This was based on our best guess and a professional analysis of the annual demand of the house. The first electrons started flowing from the system on 27 July 2012. The contractors were using electricity that we were generating to build the house as we were starting to build credit![20] Because we also

Figure 149 - South face of RISE with abundant windows and the initial 6.345 kW solar.

qualify for SREC (Solar Renewable Energy Credits), the program has paid just enough to cover our modest connection fees. A Level II electric car charger was also installed in the garage for when the right car for us becomes available.

We moved in 29 November 2012 and started learning how to best operate the house. The heat pump/mini-split system worked well, but we found that it does use a lot of electricity. We started using the masonry wood stove, and this helped offset the electrical drain. Opening southern window blinds during the day and closing them at night to trap the heat energy in the house maximized solar gain and heat retention. These efforts helped, but it seemed that we were still using too much electricity. Our kWh credit from the summer was quickly draining away. We bottomed out at 450 kWh starting out from a high of 1,800 kWh. We used ~1,300 kWh of credit plus 1,000 kWh of solar-generated electricity to get through our first winter--the critical three months of high electrical demand and short daylight hours.

It was at this point that we were honored by having HOME POWER magazine do a photo shoot and interview us for a future feature! We were surprised to be on the cover!

Our first summer approached and the overhang above the windows started shadowing the windows as the sun climbed higher in the sky. By the summer solstice, the entire southern wall was in shade at solar noon. The passive solar heating and cooling design works!

At the end of one year, the calculations showed we could have broken even by using 24 solar panels. We had three panels more than our requirements. We were over producing at this point and "banking" the extra power with the power company. This was great, but we still had plans of acquiring some type of electric (EV) or plug-in hybrid (PHEV) car in the future, and this electrical production will fall short when we do.

In February 2015, we expanded our solar array by 11 X 280-watt panels and brought the total size up to 9.425 kW. We will be over producing by a great amount and adding to our credit in preparation for our EV or PHEV.

Figure 150 - The expanded 9,425kW solar array.

In 2016, we added a much darker tile floor to the concrete slab. This absorbs much more sunlight, capturing heat and storing it away to radiate later. This made a huge difference in the performance of the house over the winter. During the winter of 2016/2017, we ran the heat pump a total of 48 hours. Between the passive solar design and the wood heating, the house was kept in 70 – 78°F range day and night. We survived through the winter months without pulling down our kWh credit.

Figure 151 is a graph of our solar production, where the blue line depicts the daily average, then starting in 2014, the weekly average in kWh production, and the red line depicts our kWh credit. The annual oscillation is seen as summer months provide long, high-sun-angle production days and winter months provide low production and higher consumption. The greater production, after the expansion in February 2015, is seen when peak output increases from ~40 kWh to ~60 kWh/day. The all-electric Chevy Bolt was purchased in February 2017. After the addition of the Bolt to the mix, we only built two MWh of credit over the summer as opposed to five MWh the previous summer.

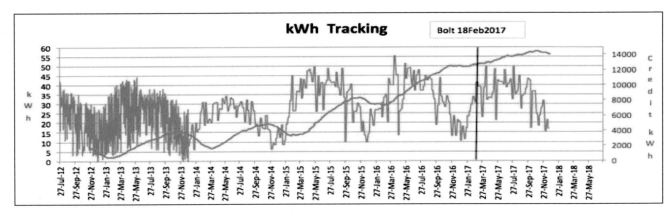

Figure 151 - Solar production and credit tracking.

We use Enphase micro inverters, a small inverter attached to each solar panel. This makes wiring simple, as the output is standard 220V household wiring. The Enphase inverters also allow one to easily track production by each panel, allowing one to easily spot any possible problem and to track daily production. One can see the production jump in 2015 when we added 11 more panels.

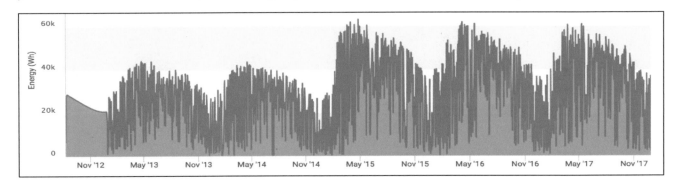

Figure 152 - Daily production over the system's life. This shows 2 1/2 years on the original system and 3 on the expanded system.

Electric Car

In February 2017, we bought a Chevy Bolt EV (indicated in the graph above). In our first year, we drove 25,046 miles and still managed to produce a surplus of 995 kWh. If you assume we replaced a 40-mpg car, our Bolt saved us over $1500 in gasoline and, more importantly, 15,000 lbs. of CO_2 not emitted into the atmosphere. We still generate slightly more electricity than we use. However, in the winter months when generation is lower and the car is using

Figure 153 - The Chevrolet Bolt EV hooked up to the charger.

more energy on the daily trip, our credit drops slightly. Yet, we achieve net zero electrical usage and emission-free driving! We also have a Subaru AWD, but the Bolt, with its 238-mile range, covers over 80% of the miles we drive. The Subaru is our only remaining use of fossil fuels for transportation until an appropriate AWD EV becomes available.

Future

We plan to expand the existing fruit orchard and vegetable garden, as well as adding a small Christmas tree farm. These efforts will help in carbon sequestration. With this addition, and our solar array likely over-producing into the future, we hope to be carbon negative. Thus far, we are very pleased with our choices, as we feel we have a bright and cheerful home to live in, we feel in control of our energy future, and believe we are leaving a clean legacy for future generations.

Figure 154 - View from loft looking down to the living, dining, and kitchen areas.

Chapter 10 – Growing Greener   by Victor Olgyay, AIA

Humans are adaptable creatures. We'll stubbornly walk with a stone in our shoe but are happier when we remove it. That certainly should be true with our homes. Rather than change ourselves to fit the home we get, we can adapt our homes to fit us better. Like a shoe, they can stretch to provide comfort, rather than force our feet to fit.

For my wife, Kristy, and me, environmental degradation has always been a stone in our shoes that we wanted to

Figure 155 - The west elevation, under the hackberry tree.

remove. Environmental issues are by their nature systemic, and it is not only the stone that matters, but also the foot in the shoe, the behavior of the walker, the ground we walk on. Solutions to environmental issues are also often interdisciplinary and multifaceted, not simple, and singular. Interestingly, the systematic characteristic also has many benefits; for example, often an individual effort will effect change in a broad neighborhood.

Reducing our environmental footprint does not happen overnight; it's a long process with a lot of individual events. Ecological responsibility is a deeply held value in our family, and we truly enjoy each move towards making the world better. Awed by the Rocky Mountains, we are respectful of the abundance of nature's resources that provide for our human needs. We want clean water, fresh air, and vibrant health in our family and in nature. In living our lives, we strive to be a constructive part of the larger ecology.

Quick Facts:
> **Net Energy Emissions** – 7050 lbs. - 83% less than typical American family
>   (131% less with offsets)
> **Home** – Renovated 1955 ranch, expanded to 2300 sq. ft., Boulder Colorado, USA
> **Technologies employed** – Passive solar, superinsulation, PV, SHW, plug-in hybrid vehicle
> **Annual Energy Costs** – $962, 79% reduction from local average

Beginning

In 2004, we had our second daughter and needed more room for our small family. We looked at a lot of properties and bought a small 1950's-era ranch conveniently located in central Boulder, Colorado. As a one car (VW Jetta) family, this helped a lot. We walk or bike to work and shopping, and meet many of our needs locally, minimizing our need to drive. Choosing this location greatly reduced our environmental impact. This location has an "Above Average Walkable" rating according to the EPA "Walkability Index", and we have several bus

Figure 156 - Goose Creek multi use path is my daily bicycle commute.

lines and bike paths within ¼ mile of our house. While still a relatively low density, suburban neighborhood, the inverse relationship between walkability and transport-related emissions is notable. Not to mention that riding my bicycle to work is a wonderful way to start and end my day.

Being an older neighborhood, our lot has a beautiful, mature hackberry tree on which we immediately hung a swing. We built a tree house for our kids in an old apple tree, planted a few more fruit trees, and started a 400-square-foot vegetable garden. Colorado is a dry climate, so we reworked our landscape to reduce water needs, planted native plants, and removed grass we did not use. We let some of our yard go wild and kept a small patch of lawn, which we mow with a hand mower. Because we have a small (6,000 square foot) lot--about half of which is covered with the house, sidewalks, and driveway--we wanted to restore the remaining arable land to be ecologically healthy and productive.

Figure 157 - Street view in 2004 when we purchased the home.

Figure 158 - Street view in 2017 after we remodeled.

The little old house was cozy, and as we lived in it, we gradually made it more so. We got a truck full of cellulose, insulated every exterior wall cavity, and put 16" of fluffy insulation in the attic. Our house sits on a crawlspace, so I spent a few weekends wrapping all the hot air ducts in the crawlspace with insulation, insulated the hot water pipes, and finally, sprayed Iceynene foam insulation between the floor joists, insulating and sealing the floor from underneath. We caulked and weather-stripped all our windows and doors. There is still a fireplace flue and a kitchen hood vent, but otherwise the house was reasonably airtight. Finally, in 2007 we bought a small 4 kW solar PV system that met most of our electricity needs over the course of a year. Sometimes we got itty-bitty annual electricity rebate checks from Xcel, our utility, and we typically spent a couple hundred dollars per year on natural gas. Our energy use was low, and the house felt like a good fit.

Winds of change

Boulder has some of the highest peak winds of any city in the US, hitting 60-90 miles per hour a few times per year. After one particularly windy evening, I was walking around our property, looking at the damage--broken branches, overturned garbage cans, and so forth. My neighbor, who was also milling about, yelled over to me "Hey! Have you seen my porch umbrella?" I shook my head no, and he then pointed to my roof. There on my roof, neatly folded up next to my solar collectors, was my neighbor's porch umbrella.

I immediately set about retrieving the umbrella for him, setting a ladder against the eve of my roof and climbing up. Little did I know my two daughters, then ages 8 and 4, were also following me up onto the roof. We sat there for a few minutes, enjoying the view, and my older daughter declared, "This is the best room in our house." And so, the next phase began.

Our 1200 sq. ft. house had begun to feel a bit cramped, and we all agreed that adding a second floor would give us more room and perhaps provide those captivating mountain views. We built a cardboard model of our existing house, and then we all participated in talking through what we wanted in the remodel. With a little help from my family and friends, I pulled the design together in 2011, and we started construction.

Boulder has a very interesting "high alpine" climate. Located at 5400 feet above sea level, the mountain air holds very little moisture, so we average 300 sunny days a year and large diurnal temperature swings – a 40-degree difference between day and night is not unusual! While the dry climate is very nice in many ways, it also results in low rainfall, around 20 inches/year. This makes water a precious resource and determines a lot of the local ecology. Boulder can be hot in the summer and cold in the winter, (winter design temperature 4°F, summer 89°F), so a house designed for this climate needs to adjust to these varying conditions.

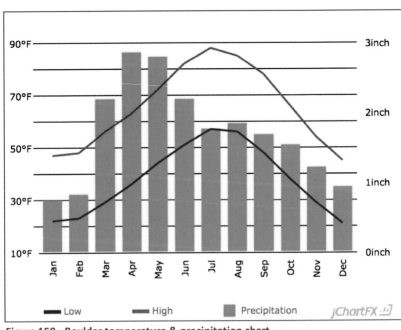

Figure 159 - Boulder temperature & precipitation chart.

Overlaying Boulder's climate data on a psychrometric chart shows that there are a lot of hours during the year when the Boulder climate is comfortable, a lot of hours when heating is necessary, and a few when cooling is desirable. Given these conditions, many typical bioclimatic strategies will produce good results.

We designed the remodeled house to be very well insulated and have generous south-facing windows with overhangs for summer shading. We

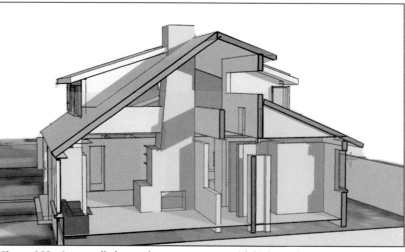

Figure 160 - A centrally located, two story area with high windows letting winter sunlight reach the north side of the house and venting hot summer air.

created a centrally located, two-story living room space to assist with natural ventilation in the summer and provide bilateral daylighting in most rooms. We also added a greenhouse that helps heat the house as well as grow plants.

We added a 900-square-foot second floor and reclaimed 200 square feet of garage space, nearly doubling the square footage from 1,200 to 2,300. Since we were tearing up the place, it was an ideal time to do a few other performance upgrades. We switched out the existing double-pane, low-e windows (with terrible frames) for quad-pane, fiberglass-frame windows with an overall unit insulative average of R9. We decreased our home's north-facing glass and increased it on the sunny south side. We calculated the overhangs to provide passive solar heating in the winter and shading in the summer. All the new windows are casements or awnings, so they lock airtight when we want them to and catch the wind when open for increased ventilation.

And we did a million other things. We reused as much of the old roof framing as possible, framing the upstairs walls with the old 2x8 rafters, stuffing them with cellulose, and wrapping the whole house with two inches of rigid insulation to eliminate thermal bridges. The roof is framed with R40 SIPs. We used an ENERGY Star-rated light-colored roofing material, and locally sourced, beetle-kill pine for the soffits and trim.

The house performs well, using about one-third the gas and one fifth the electricity of a typical house this size. But most importantly, it's a comfortable fit. When it is 95 degrees Fahrenheit outside in midsummer, inside the house is 76 ... and we have no AC. We take advantage of Boulder's diurnal temperature swings and open the windows at night to let the cool night air in and close them during the heat of the day.

Figure 161 – Correctly-sized overhangs on the south side of the house keep the summer sun out with passive solar design.

Yet the true comfort is in how we use the house and how it supports our lives. Our kids love the balcony over the living room; they can spy on the adults or fly airplanes down on unsuspecting targets. The community spaces—living room, play room, deck—gather us as a family, while private spaces let us take quiet time. We built 1-watt LED lights into the stair risers to cast an amber glow on the adjacent wall, safely and satisfyingly identifying the steps. Their warm, low color temperature and low light levels have the least amount of impact on sleep cycles and the body's melatonin production, allowing people to traverse at night without triggering their brains to wake up any more than necessary.

Figure 162 - Low color temperature LED lights illuminate stairs.

The west side of the house, nestled under the shade of the big hackberry tree, is lousy for solar collection, but a great place to plant a green roof. We planted native grasses and sedums, creating another 600 square feet of habitat for birds, squirrels, and butterflies. Green roofs are a great way to help keep the house cool in the summer, yet warm in the winter, while still providing habitat. With a view of the Rocky Mountains' foothills, this is the "room" that my neighbor's umbrella revealed to us. Architecture must support life first.

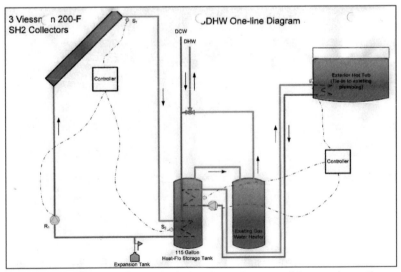

Figure 163 - Our planted roof, gradually growing greener.

The rest of the roof collects rain and sun. We pulled off the existing 4 kW PV system during the renovation and reinstalled it afterwards. We are effectively net zero for electricity—the average American home consumes ~10,800 kWh of electricity per year; between July 2014 and June 2015, we consumed just 96 kWh from our utility. And because our house is significantly more airtight than it used to be, we decided to get rid of combustion within the building envelope. No fireplace, no gas water heater, no carbon monoxide. We have an inexpensive electric water heater as back-up, but most of our domestic hot

Figure 164 - A diagram of our solar hot water system, supplying our domestic needs, including the hot tub.

water comes from the sun. We added a solar thermal domestic hot water system with three collectors yielding 90 kBTU/day average--enough for most of our needs in the winter and more than enough in the summer. We store the extra heat in our hot tub.

We still have a natural gas furnace, albeit in the crawlspace outside of our building envelope. We plan to replace it with an electric, air-source heat pump. However, until then we have been purchasing 1000 kWh/month blocks of wind power to offset this carbon source.

And the road continues

This year (2017) we bought an Audi A3 e-tron, a plug-in hybrid 4-person car which is rated at 86 mpg equivalent. It's "electricity only" range is low, only 20 to 30 miles, but because a lot of our typical daily trips are just around town, we have many days that we only use electricity, sourced primarily from our home PV system. We have driven about 6,500 miles in the first 6 months of owning this vehicle, so we are anticipating logging about 14,000 miles per year, about 60% of which is on electricity only.

We still have much to do to reduce our family carbon footprint. We recently worked with "Snugg Home" to outline a path to get to a zero-carbon goal with a reasonable return. Snugg--a City-of-Boulder sponsored program--takes an innovative systems approach to encourage people to reduce energy use. Because of the ROI associated with energy efficiency and electric vehicles, loan financing can be essentially paid for through energy cost savings, so there is zero net cost to the homeowner for these energy improvements.

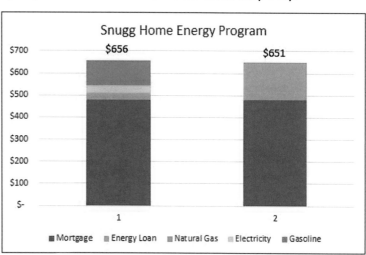

Figure 165 – Same monthly payment with a lower carbon footprint! – Lower energy costs offset the increased finance costs with energy savings.

In our example package, we identified several needs that increased efficiency, reduced carbon, or increased comfort. This included such measures as replacing our existing 92% efficient gas furnace with an electric air-source heat pump (ASHP), buying a new, efficient refrigerator, adding an air-to-air heat exchanger for more fresh air, and increasing the size of our PV system to 8 kW to take care of our additional electric loads from the ASHP and our hybrid car. In our 2016 analysis, this paid off in about 8 years, saved 10 tons of carbon, and in addition would save me $7,400 over 10 years. Part of the genius of this approach is that several energy measures are bundled together. In this way, measures with a quick payback (lighting, refrigerators) subsidize longer-term items (like windows), creating an overall acceptable payback. We are in the process of implementing these ideas. In my case, we should offset $177 in energy costs with a $172 energy loan, a savings of $5/month, so we never see the increased costs used to finance the energy-saving projects.

Ultimately, the impact of our house is less about the dollars we save--or even the energy. What we really want to know is have we had a positive or negative net environmental impact? We get most of our ecosystem services, (like clean air, carbon sequestration, pollination) from biological systems, which have metabolisms and thus measure their benefit over time. It makes sense that we also consider our carbon production or reduction over time and include biology in the calculation.

Building a house uses a lot of energy; there is energy in creating the materials, transporting them to the site, and still more in assembling them on site. This energy (and carbon) is then "embodied" in the building, becomes a "sunk cost." Starting with an existing building takes advantage of this resource, so renovating an existing building typically has lower environmental impact than starting from scratch. A recent study by the Preservation Green Lab of the National Trust for Historic Preservation, entitled, "The Greenest Building: Quantifying the Environmental Value of Building Reuse," concludes that, when comparing buildings of equivalent size and function, building reuse almost always offers environmental savings over demolition and new construction. But even if lower, can that embodied carbon debt ever be paid off? Yes, it can, either through increased biological sequestration, and/or by over-producing clean energy to offset the embodied energy. This is here demonstrated in the "Green Footstep" calculation, which shows our house paying off the carbon debt from the renovation in 20 years.

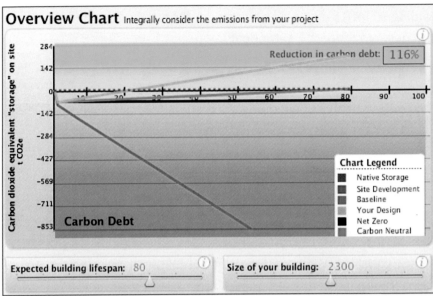

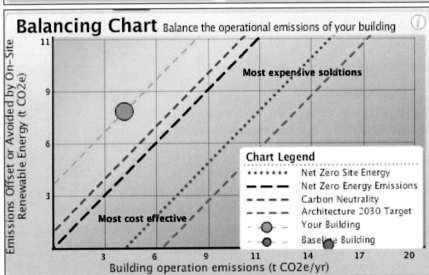

Figure 166 - Life cycle carbon assessment using Green Footstep analysis. Overview chart shows the embodied carbon debt repaid after 20 years, Balancing chart shows 4 tons CO₂e offset beyond net 0 operational emissions. Analysis includes wind power offsets.

Figure 167 - Solar PV on upper roof and solar thermal on lower.

Epilogue

There is something quaint and anachronistic about independently analyzing this house as if it were hermetically sealed under a dome. Reality is much more interesting. We are connected to our neighbors and have a shared interest in our air, our water, and the weather. Thinking beyond the property lines, I benefit when my neighbor combusts less fuel. Making our greater neighborhood sustainable is easier than

Figure 168 - The south elevation on December 21: PV, solar hot water, and sunlight streaming in.

doing so house by house; houses that produce more energy than they use can offset houses that use more than they produce. Houses with larger lots can sequester more carbon than those with smaller lots. Community resources--like solar gardens, shared transportation, or parks--are all a critical part of building a healthy ecology. Watersheds rarely follow property lines.

Adding up all our individual efforts becomes significant. We are slowly decarbonizing the electricity we get from the grid, and each house helps. We do what we can right now, recognizing that there will be more opportunities tomorrow to further economize, to cultivate our dwelling on earth.

But in the meantime, we have fruit trees, a garden, and a home that's comfortable, efficient, and

connected to nature. Our little 6,000-square-foot landscape is diverse, encouraging more pollinators to visit, birds to spread seeds, and other ecosystem services to flourish. As a family, we are adapting to environmental concerns and slowly working to improve our individual and collective lives. Our home needs to fit us, and by doing so, it also better fits the planet.

Figure 169 - The west elevation in summer, shaded by the hackberry tree. Note our planted roof.

Chapter 11 – Self-Reliant Grandeur   by Lloyd Marcum

In building our home, our main goal was self-reliance: the ability to power our home, vehicles, and well water if the central utilities were down for an extended period. Additionally, we wanted to be net 0 in electricity costs, net 0 in electrical use, net 0 in fuel costs for our cars, and net zero in property taxes. Read on for an inventive solution. Initially, pragmatic long-term cost savings drove us, but as the children grew, the environmental aspects of our decision became important to us, as well.

Quick Facts:
   Net Energy Emissions – 16,000 lbs. - 80% less than similar-sized homes in the area
   Home – 10,000 square feet in Central California, USA
   Technologies Employed – Solar PV, electric vehicles, heavy insulation
   Annual Energy Costs – $2000, down 82% from comparable neighbors

First a little background on me. By no means was I, fifteen years ago, on a course towards a Net Zero lifestyle. I was (and still am) a motorhead, always striving for more power in cars and airplanes. If someone told me that I would become an owner of a net Zero home, I would have told them they were crazy. My cars started with a 69 GTO, a series of Porsches, BMW's, and muscle cars to, finally, a Tesla Model S. Airplanes, though not electric yet, went from small Cessna's to a Beechcraft Bonanza to light jets and turboprops. I now own a small charter service and fly

Figure 170 - One of the Pilatus PC-12 planes we use in a charter business.

exclusively Pilatus PC-12's. The PC-12 is a very efficient single-engine turboprop for up to 10 passengers.

In 2004, when my wife and I decided to build a custom home, our thought was to make it smart and as efficient as possible. We purchased a 5 acre lot in an area near the central California coast. We wanted to build at the end of a cul-de-sac, properly situated for the prevailing wind. We planned to have the home and pool on the front three acres, leaving some land for wine grapes on the back section zoned for agriculture. One of the benefits of this lot is it has its own well for another level of self-sufficiency. The first order of business was to establish electrical service and complete the well by setting tanks and a pump in the well. We had to trench about 300 feet to a location where the meter panel was going to be. Note that PG&E, our electric service provider, pays for the wire from the distribution point to your meter panel. If you are going to be doing any of this work yourself, as I did, you will need to become familiar with the 'Greenbook' of codes and requirements for commercial

Figure 171 - Two 200-amp service meters.

properties. I planned to have two separate 200 ampere services for this project--one 200 amp service for the home, and one 200 amp service for the agriculture portion, which includes most of the electric vehicle charging. I ordered the appropriate panel, and constructed the panel backboard out of ¼ inch steel and pipe. Now the green book will tell you that the minimum standard for the panel is a backboard made of wood. Many homeowners follow the minimum standard, but when the wood rots, you then are required to pay the utility to disconnect you while you rebuild the backboard. PG&E was so used to seeing wood that I had to argue to several supervisors that my construction was above the standard the green book set. Steel construction assured a longer life for my panel, and, fortunately, I prevailed. PG&E pulled the wire in the conduit we provided for them.

The next order of business was to secure a place for my water tanks. I framed and poured a pad to accept two 5500-gallon water storage tanks above ground. This quantity was needed for fire protection due to the size of the home, and it covers projected usage in both the home and vineyard. Immediately following the placement of the tanks, I contacted a local solar provider and indicated that I wanted a solar system to be able to provide enough electricity to pump water from the well and provide power for construction of our home. We

Figure 173 - Well with storage tanks and solar power.

contracted with a local solar provider to construct a 3.5 kW solar array above the water tanks I already placed. This installation was expensive per watt, as the cost for panels was high in 2005. Still, this small array provided enough power to pump all the water and the power needed to construct our home over the next three years. Yes, we were three years in construction! The house includes a swimming pool and employs passive-solar pool heating.

Figure 172 - SunnyBoy inverter converts DC power from solar panels into AC power for construction tools.

We constructed this home with 2X6 exterior walls which enabled higher 'R' values for the insulation. We employed in-floor radiant heating with a separate zone for each room in the home. This allows us to turn off sections of the house when not in use and effectively heat a much smaller footprint than the actual 10,000 square foot house. We use natural gas for heating and cooking, which is our only fossil fuel use.

Upon moving in we quickly realized that without the benefit of solar the house meter was costing us much more than we had planned on. I duplicated the design of the original solar installation and constructed a ground-mounted array for an additional 3.5 kW, bringing us to 7 kW, located in the SW corner of the property.

Figure 174 – The house under construction, featuring 2x6" walls for added insulation.

Understand that we have a lot of electric demand in this 10,000 sq. ft. house. We have three teenage kids, water pumps, a pool and pumps, three refrigerator-freezers, a cooled wine cellar, a chest freezer, agricultural pumping of water for two acres of grapes and three electric ovens. My neighbor's electricity bill without solar was in the $1200 to $1500 range monthly.

Figure 175 - Overhead view of homesite with vineyard in the SE and solar arrays near the NE & SW corners.

California has very high electric rates, averaging about $.24/kWh. Yet on an annual basis, we have been able to bring this house to a Net Zero cost for electricity bills. During the winter months we drain our accrued energy credits which were built-up during the long sunny season.

The completed home and the pool are shown in Figures 176 and 177 below.

Figure 176 - Completed home with agricultural building to the right as viewed from the SW corner of our lot.

Our home was featured in a privately-financed promotional video for the Model S, which can be found in You Tube under "Gallons of Light" by Jordan Bloch. Working with the film crews was fun and informative watching them make multiple takes in the correct light for just a few seconds of the final production. It is well done and worth looking up.

Figure 177 - Our pool and pool house behind the home. The pool is heated with thermal solar and the pumps powered by solar PV.

As my kids entered high school, I found that our electricity bill continued to grow. I decided to add an additional PV array to the agricultural side, and subsequently to the house side, as our usage increased.

Now, with 20 kW of solar, we achieved a surplus in electricity production!

Figure 178 - Expansion of our solar system to 20 kW.

Agriculture Side –

Adding income to the property from farming was accomplished by adding two acres of Pinot Noir wine grapes on the agricultural side of the property. I purchased 2000 plants from a vineyard in the central valley settling on a Clone 626, which is resistant to disease and provides a good, rich yield. I started producing a commercial crop of grapes after 3 years, and now after 12 years, produce 8 tons of grapes per year, which we sell commercially for about $2500 per ton. This provides enough income to become Net Zero from California property taxes. We have zero expenses for water and electricity to run the farm, thanks to our solar array and well. We have some expenses for labor during harvest and pruning, and minor expenses for sprays and fertilizers.

Figure 179 - Pinot Noir wine grapes are growing well in the vineyard.   Figure 180 - Last year we harvested eight tons of grapes.

Transportation –

In 2012, we added two electric cars--a Tesla Performance + Model S, and a Toyota Rav4EV--to our stable. With a Tesla Model 3 on order, we should use all our 20 kW of solar array. The electric cars are my best purchase of all time. No gas bills, no electric bills and no more pumping gas, changing oil and disposing of it. Driving for only the cost of insurance and minor maintenance is wonderful. And the performance of these EV's is second to none, making it easy for me to move away from my muscle cars and the Porsche's I used to drive.

Figure 181 - Tesla Performance + Model S and saying good bye to oil.

Figure 182 - Tesla Model S and Toyota Rav4EV parked in front of our expanded solar array, now at 20 kW.

Backup Power

I was feeling a little vulnerable to the electric company, since we are grid-tied and wanted backup in case of an emergency. I found a good buy on a used 30 kW generator and made a project of constructing a backup power supply. I know this burns diesel fuel, but it is used rarely--only in an emergency. This has the capacity to power all the electrical needs. Some will ask why I did not employ batteries or a Tesla Powerwall as my backup, and the answer is simply they cost too much, at this time, to supply that much power. I would need six Powerwalls to generate 30 kW of power at a cost of about $35,000 plus installation.

Figure 183 – Our 30 kW generator for emergency backup.

Conclusion

Overall, we are pleased with our outcome. We targeted net 0 in electricity costs, net 0 in electrical use, net 0 in fuel costs for our cars, and net zero in property taxes, all of which we have achieved. Our only "adjustment" was that last year I had about $2300 of surplus electricity that went back to PG&E. When you don't use what you have produced at an average of about $.24 per kWh, PG&E gives you back a token $0.04--the wholesale rate!

Therefore, I decided to find a use for the surplus electricity we are currently producing. Bitcoin mining was the answer. After purchasing some computer mining equipment, I now use the excess power that previously went back to PG&E at little benefit to us. Time will tell if this is a viable operation. The first month of operation I produced about $1800 worth of Bitcoin on "free" electricity. So, I am hopeful!   But if this does not work out, then I am confident my rising teenage drivers will put that excess production to good use.

Figure 184 - Bitcoin mining, a handy use of our current power surplus.

Chapter 12 – Long View – A Canadian Perspective   by Richard Corley

Our path toward net zero living goes back more than 35 years, when we decided to try to make our first family home a zero-carbon one. We were in our early 20s at the time and certainly couldn't afford to embark on any strange science experiments but wanted to try to minimize the impact of our housing on the planet, including by avoiding the use of fossil fuels in our house. This decision was considered a bit radical at the time, as we were building in a neighbourhood that was served by natural gas, and every other house was connected to and relies upon natural gas for heating.

In reality, our path to net zero began much earlier than that. Our house reflected a continuation of work that I had begun in the 1970's, in high school, when I converted a utility van into a mini motor-home, which stored and recycled waste engine heat to provide hot water and space-heating and used three lead-acid batteries and homemade power inverters to provide AC power for a refrigerator and other electrical appliances. I was always interested in making more efficient use of energy. At a deeper philosophical level, underpinning it all is a sense that waste is morally, as well as economically, wrong.

Quick Facts:
    Net Energy Emissions – 0 lbs. of CO_2 from energy use, 100% less than typical households.
    Home – 5,000 square feet in Ontario Canada
    Technologies Employed – Passive & PV solar, heat pumps, EVs--all energy from green electricity
    Annual Net Energy Costs – $0, down 100%. Solar revenues exceed electricity charges.

Figure 185 - Drone view of east side of home, with solar array on south-facing roof and garage acting as air lock & buffer on north side.

How We Got to Net Zero

We started by building a low energy house, which used solar energy and an air-source heat pump, instead of natural gas, for space heating. We increasingly powered the house with low-carbon electricity as it became available. Ontario has one of the cleaner grids in North America, with around 90% of electric production CO_2 free, and we pay Bullfrog Power, a green energy provider,[21] to inject renewable energy into the grid to fully cover all the power we consume. In addition, in 2010 we added 12 kW of solar panels to the entire house to power the house and to feed power back to the grid. In 2012 we purchased one of the first Tesla Model S in Canada and in 2016 replaced our Prius with a second Model S. The electrical energy produced by our own solar panels, together with the community solar projects that we supported, provide a multiple of our annual power consumption. We travel by electric vehicle wherever feasible, including trips from Canada to Florida. Low-carbon sports include cycling, kayaking, wind surfing and cross-country skiing. A vegetarian diet, home vegetable garden, and preferential consumption of local organic produce round out the picture of our efforts to get to or beyond net zero.

We will discuss, in turn each our efforts to cut our carbon footprint.

Low Energy House – The least expensive energy is typically that which not required or consumed. Our house was designed to minimize energy consumption and loss through both conduction and air infiltration and to recover as much of the energy from exhaust air as possible.

The use of only electricity to power the house was a key design element, which enabled many energy conservation features. The decision to eschew the combustion of natural gas (or oil) for heating, cooking, hot water, and the drying of clothes made possible an air-tight building envelope and the recovery of heat energy associated with the performance of these functions. We designed the house to collect and channel the warmth and/or moisture from washrooms, cooking, drying of clothes, central vacuum, air compressors, and solar power inverters to a central heat exchange room, where the heat and latent heat of evaporation is recovered in two stages. First, a General Electric heat pump water heater extracts energy and condenses water vapour and transfers the

Figure 186 - GE heat pump water heater, 3X more efficient.

heat to the hot water tank. It provides around three watts of heat energy for each watt of electrical energy it consumes.

Second, a high-efficiency air exchanger from VanEE recovers and transfers the heat recovered from the outgoing exhaust air into the incoming fresh air stream, including the latent heat of evaporation of the water vapour in the exhaust air stream.

Other low energy design features included a 100% continuous vapour barrier (including over the ends of floor joists, behind the masonry fireplaces, throughout the basement, and connected to all window and door

Figure 187 - Air to air heat exchanger keeps the air fresh with minimal heat loss.

frames), the use of flexible seals around all windows and doors, high R-value insulation (R60 in roof, R30 to R40 in walls, and R20 below all floors), and thermal breaks between internal and external masonry.

The house was designed to create air locks between areas at the north end of the house to reduce the energy lost when entering and exiting the house. The garage, which is well insulated but not heated, and has insulated doors and storm doors, is relatively air-tight and is warmer than the outside temperature in winter. It serves as the first air lock. The rear entrance room / workshop, which is part of the house but has the heat vent closed, serves as the second air lock. It is warmer than the garage but colder than the rest of the house. The front hall coatroom area and entrance hall further buffer the entrance areas from the rest of the house.

Figure 188 - Solar inverters in the central heat exchange room.

Thermal Solar Energy
(Passive Solar for those south of Canada)

Effective use of solar energy for residential space heating purposes in a temperate climate requires three things: access to sunlight when heat is required, the ability to exclude the sunlight when it is not required, and the capacity to absorb and store the energy received when the sun is shining to prevent overheating, then enable the use of the stored energy after the sun goes down.

The solar energy is received through south-facing windows and skylights. We have a total of six south-facing windows and six south-facing skylights. The window units average roughly eight feet by six feet and the skylights are four by four feet. These double-glazed units provide substantial winter thermal energy when the sun is shining.

The sunlight is controlled and, when not required, excluded by the combination of an appropriate overhang and external roll shutters. At our latitude in southern Ontario, we determined that a one-in-three ratio of overhang to vertical height provides energy input, which, on sunny days closely match our thermal energy requirements throughout the year. The overhang completely excludes sunlight from the windows for the months around the summer solstice, while allowing increasing amounts of sunshine to enter as the winter solstice approaches. This compromise provided more energy than required in the autumn, but less than is required in the spring. Rollshutters on the skylights and windows enable exclusion of the sunlight from the skylights through the summer and of unwanted energy that would otherwise be received through the windows on hazy days in the summer.

Thermal Energy Storage

The building materials used in the house provide the thermal storage required to make effective use of passive solar energy in the winter and to largely avoid the need for air conditioning in the summer. They also permit the effective use of electricity to heat or cool the house when the price of electricity is low. Concrete, masonry, tiles, drywall, and other materials--all of which are insulated from both the outside air and the earth--provide substantial energy storage capacity. The main floor on the south side of the house consists of steel reinforced concrete, which is topped by tile and absorbs energy from direct sunlight during the day. The forced air system is then used in fan-only mode to circulate the warm air from the south side of the house to the other rooms. The masonry fireplaces and the tiled concrete floors in the basement are insulated to at least R20. The insulation under the tiled concrete floors also increases the level of comfort and makes the basement level rooms more comfortable, while also enabling thermal energy storage.

The thermal storage, together with the control of solar heating enables the house to be kept comfortably cool throughout summer with minimum (if any) air conditioning. In most cases, the nighttime temperature is cooler than the temperature inside the house, such that the house can be cooled by venting the warm air and replacing it with cool outside air which is circulated using the furnace fan, thus reducing the heat accumulated during the day from the house's thermal energy storage. In 2017 the house has remained comfortable--between 68 and 76 degrees Fahrenheit--from May to October with no need for either heating or air conditioning.

Thermal and Electric Energy Management

Electrical power rates in Ontario vary with the time of day. Therefore, thermal energy storage allows the house to completely avoid the use of any electricity for heating or cooling during the times of peak electricity prices. We use an Energate internet connected thermostat to shift the use of electricity to heat or cool the house and to heat water to periods when the electrical power is in ample supply (and prices are low).

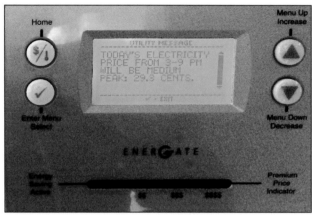

Figure 189 - Energate internet enabled thermostat.

Air Source Heat Pumps

Even though we have cold winters--January and February average 28°F (or -3°C)--we have used heat pumps with success. Our Hallowell heat pump was designed in Maine for very cold temperatures and has been used for the past 10 years. Unfortunately, the company is now out of business, and we are about to replace that heat pump with another high-efficiency, cold-weather heat pump. Special cold climate heat pumps have a COP of 2 at 0°F (-18°C), meaning they are still twice as efficient as resistance heating. This allows us to remain combustion free while keeping the house comfortable, thus keeping energy usage to a minimum.

Figure 190 - Cold climate heat pump is good down to 0F.

Solar Panels

We have 54 SunPower solar panels for a total system size of 12 kW. While on the large size for most home systems, remember that eastern Canada is not noted for its sunny weather, so we compensated with a larger array. We average about three hours of peak sun a day--well below the 5-6 hours one can harvest in the American SW. This system generates approximately 15,000 kWh per year and is grid tied. Additionally, we help support community solar in our area through a project of the

Figure 191 - 12 kW solar array, powers the house and cars, installed in 2010

Unitarian Community Church and SolarShare Community Power Coop. SolarShare has grown to be North America's largest solar cooperative, with 500 members and has over six MW of solar generation.

Electric Vehicles

In 2012, we purchased one of the first Tesla Model S sold in Canada. Later, we upgraded our Prius with a second Tesla, a 2016 90D Model S, which has all-wheel drive to help in the Canadian winter. As a bonus, the all-wheel drive goes farther on a charge. To avoid any possible contention over access to charging, we added a 100-amp subpanel to support two Tesla HPWCs in our garage   The Tesla HPWCs share a 100-amp circuit allowing each car to charge at 40 amps or 80 amps for one car, but we typically charge the cars overnight at 20 Amps (5 kW) to minimize peak demand on the distribution system and to keep warming the battery pack overnight.

Figure 192 - Two Tesla charging stations on a 100-amp subpanel.

Figure 193 – Our two Teslas charging in the garage.

It is so convenient to wake up each morning with a full charge and not worry about getting out to pump gasoline. We can even set the cars to precondition (warm up) before we are ready to go out, so the car is warm when we get in, without ever burning any gas or creating any fumes in the attached garage. Overall, they have performed beautifully here in the great white north, adding convenience and saving us money at the same time.

I had the opportunity to test a hydrogen fuel cell vehicle. And while they are cleaner than a gasoline powered car, you still need to find and then go to a filling station to get more hydrogen--far more inconvenient and expensive than charging at home--especially considering we produce the power ourselves from the rooftop solar. Additionally, it is very hard to beat the performance, space, and other features of the Tesla Model S.

Figure 194 - Testing a hydrogen fuel cell car with home in background.

Diet

Our diet is vegetarian, and we plant a home vegetable garden each year to supply some of our food needs. Beyond that, we prefer consumption of local organic produce. This helps to round out the picture of our efforts to get to or beyond net zero.

Low Carbon Travel and Recreation

When possible, we do our traveling with our EV, which is powered by solar at home and with lower-carbon electricity through the Tesla Supercharger network when away from home. We have just completed our fourth annual Tesla-powered driving vacation to Florida, which is getting easier each time with the continuing growth of the Supercharger network. The carbon footprint of EV travel is a small fraction of taking a plane.

Figure 195 - Travel in an EV is lower carbon than plane travel for us to Florida.

For recreation we prefer lower carbon activities such as cycling, kayaking, wind surfing, cross-country skiing, and biking in the snow.

Many of these activities we now enjoy with our grandson, who is a major reason for our net zero journey. The less waste we generate, the more resources we leave for future generations to use and enjoy.

Figure 196 - Winter tires on the Tesla and bike allow for year-round fun.

Figure 197 - Bicycling near home with our grandson.

Figure 198 - Teaching our grandson to kayak.

"Society grows great when old men plant trees whose shade they know they shall never sit in."

— Anonymous Greek Proverb

Chapter 13 - The Radical Moderate   By Mark Smolinski

The Angry Man - On a pleasant day in October 2016, I took my daughter and grandchildren for a ride to see the wonderful fall foliage in the Adirondack Mountains of New York State. We pulled over in the little hamlet of Inlet, NY and I popped open the falcon wing doors of the Tesla Model X, so the grandkids could go check out the beautiful view of the lake. I was immediately approached by a small group of older ladies and was asked if I was driving a DeLorean. After a pleasant conversation about my electric vehicle and Supercharging, I noted an older couple standing right behind these women. The man was visibly shaking. I soon learned he was shaking with rage.

When the ladies left, he stepped up and sarcastically asked, "But where does it get its power from?" It was clear I was about to get a lecture from someone who had listened to too much right-wing propaganda about power plant emissions. I told him that when I am in Florida, I drive around on sunlight from my solar panels, and when I am in New York, I buy all-renewable electricity, which is hardly any more expensive. He became even more enraged and started screaming at me that I was apparently lying about my electric bill in New York State. At that point, his wife pulled him away.

This story illustrates an important issue for those of us who are working to improve and protect our environment by moving toward net zero: Not everyone agrees about the need to do so or supports those of us who do. In fact, for some, what we are doing appears to them a threat and arouses anger and hostility. That is the reality of the times and the political climate in which we live. Part of our work is to recognize this fact and do our best to educate and enlighten those who oppose our efforts. Keep in mind that underneath most anger is fear. We must also accept the fact that, despite our best efforts, some will never listen or change. Our goal should be to lead by example and share our knowledge and experience with those who will.

Quick Facts:
   Net Energy Emissions – 1500 lbs. - 97% below average
   Homes – 2,600 sq. ft. in New York & 3,400 sq. ft in Florida, USA
   Technologies Employed – Solar PV, electric vehicles, rain catchment, low water grounds
   Annual Energy Costs - $750

The Born Pragmatist - My journey to net zero probably started at birth. From an early age, I have loathed waste and admired—and pursued--efficiency. When I began driving, I realized that city driving was really just a race to the next red light. Nobody wins. I learned to take my foot off the gas when the light ahead turned red. Why waste the gas? The first brake job on my Buick Roadmaster came at 110,000 miles. Saving gas also saved on maintenance costs!

My distaste for two-seater sports cars probably started the day I couldn't go for a ride in my father's Corvette because he already had one passenger going with him. Why drive around with only one seat and no trunk? What a waste! Thus, my first car out of college, a Caprice Classic, had an engine powerful enough to tow things behind it. The trunk in my Roadmaster was large enough to hold the two deer I bagged on opening day in 2001, and it was large enough to hold my deflated 10' dinghy and its 6 HP outboard motor.

I drove like an old man my entire life--until I started driving electric, with power supplied by the sun. When you are burning nothing but photons, when braking is putting energy back into the battery, and when you have maximum torque available from a standstill, you discover you can drive with a heavy foot and really enjoy the ride. I won't say that I now drive like a lunatic...but my wife might.

Figure 199 - Our current set of electric vehicles, inherited their tags 'SUN FUEL' and 'OVR OIL' from the 2012 Volts we originally purchased to 'drive on sunlight'.

Free Water from the Sky - If there is a defined road to net zero for me, mile marker zero would probably be the installation of cisterns during the construction of our Florida home in 1996. However, it wasn't about carbon, ecology or any other 'green' thought. It goes back to that part of my character that hates waste. Water bills in Florida are expensive, and the seemingly unending urban growth leads counties to sue each other over potable water access. After I read an article about a plumber who installed cisterns, reminding folks how much free water fell on the average roof every year, I contracted him to build us a

Figure 200 - Solar still condensate collection system.

system with five 1500-gallon underground tanks for 7500 gallons of the best water money doesn't have to buy. I had no idea how far-reaching and beneficial this decision would be. Once you consume nothing but rainwater, you note how much chlorine you had previously been consuming. Chlorine kills yeast in bread dough, bleaches out hair dye, and makes home-made soda taste awful. In the last twenty years, we have also learned that drinking-water sources are now tainted with the pharmaceuticals we excrete, and even Florida's aquifers are becoming polluted with runoff from surface sources. On top of the benefit of free, quality water coming onto your home, connecting your gutters to storage tanks results in fewer drainage problems around your property and reduces the burden on the public drainage systems. Widespread cistern use could mitigate some flash flooding and runoff pollution issues in urban areas.

While we have never cut the cord to the city water line, we rarely use city water, with months or years going by before the occasional need to top off the pool in the dry season. By using little city water, we reduce our carbon footprint in terms of the energy needed to pump water to our home. In addition, we are much more aware of our water usage. When you start to think of that water in your tanks as a finite resource, you realize that things like St. Augustine grass are an unsustainable drain on resources. Add to that the environmental impact of fertilizer and pesticide runoff, and your worldview changes.

Driving to Net Zero - As a society, we are hopelessly, endlessly spoiled and entitled. Part of the reason we have to purposefully think about 'green' issues is because we are disconnected from the streams of our supply and our waste. We open the spigot, flush the toilet, put the trash can by the curb and flip the electric switch. We hop into the two-seater sports car and make a quick jaunt to the grocery store for those couple of items for tonight's dinner, assured that the grocery store will continue to have neatly packaged foodstuffs every time we desire. We lose track that a living creature's life was ended so that we could enjoy our roast beef sandwich, and we lose track that that living creature had its own food and waste streams that needed to be managed.

My journey to net zero is motivated by practical thoughts about waste and efficiency. I doubt that the fellow who displayed anger over my Tesla would take that view if he had to shovel, quite literally, the ton of coal necessary to meet his monthly electricity needs[22]. Maybe, if he saw the coal ash left over and had to find a place for it, month after unending month, he would feel differently, especially after he learned that the coal ash has mercury in it that, unlike his fireplace ashes, would poison his garden. If he compared the labor required to do a once-and-done solar panel installation versus a never-ending coal-handling job, he would see how efficient and smart it is to drive a solar-powered car.

My grand old 1993 Buick Roadmaster was a bridge from the past, as well as the impetus toward the future. It could carry six people, tow a two-ton boat, and carry large suitcases or a smelly deer carcass (or two) in the trunk. When I got the car, I was still on active duty and my kids were in elementary school. It saw the graduation of my wife, Donna, from dental school, her scratch start-up of a dental practice, as well as the relocation to our purpose-built office for that practice. I drove it to three quarters of the nation's dental schools, lecturing on running a dental practice and marketing my software. By the time it left, only really needing a minor repair, it had 140,000 miles, was 15 years old, and my second grandchild was on the way.

My first hybrid was not bought because of ideology; it was bought because I wanted a replacement for the utility of that Roadmaster...but without the crappy gas mileage that most other replacement options offered. I was shocked to find most vehicles that could haul stuff and people had gotten so pathetically inefficient with gasoline during my tenure with the Roadmaster. Sedans were no longer made that did what I needed; I would be stuck with a Sport Utility Vehicle. So, when I did a search for fuel efficiency and SUVs, I was directed toward these newfangled 'hybrids'. Investigation taught me about the batteries, regenerative braking and the engine shutting off at slow speeds. Wow! I was the kind of guy who would throw the engine in neutral at traffic lights...why did it take them this long?

Our first hybrid was a 2006 Toyota Highlander. If you hate waste and love efficiency, you quickly become addicted to having the engine turn off at stop lights, and you look around in amazement that nobody else is becoming indignant that their engines are burning gas and needlessly stinking up the air. And that 2006 Highlander came before fracking and the fall in gas prices. Gas prices kept going up and up. The wisdom of a hybrid kept making more sense. Our 8-cylinder, 2001 Aurora was replaced with a 2008 Prius. That same year, we got our Tahoe Hybrid.

Not long after we bought our Prius, I read about folks who were paying to have batteries put in their vehicles that they could plug in and leave the house with a full charge of electricity. This was absolutely a fundamental change for me--an epiphany, of sorts. I started thinking about the big picture as I never had before. The question became, 'WHY NOT?' And I exclaim this in capital letters, because I started thinking about the symbiotic concepts of energy and lifestyle as I never really had in my lifetime. WHY NOT leave the house with a full battery? WHY NOT just stop burning fossil fuels altogether? WHY NOT get solar panels and live my life in an emission-free way? Between the cost of petroleum and its environmental impact, solar energy was now making so much sense.

Thus, began the cascade of changes. I quickly ruled out an after-market alteration of the Prius because no regenerative braking energy made its way into those batteries. Instead, I learned of the upcoming Chevrolet Volt, with its approximately 40 miles of exclusive battery use, followed by a gasoline generator backup. While I waited through the

Figure 201 – 5 kW solar system installed on the Florida home.

development of the Volt, I solarized everything I could. I started with a 5 kW photovoltaic (PV), net-metered solar array on the home roof, shown here. In those days, the convention required a uniform array on a singular roof plane to be connected to a single inverter.

Note the 3 panels farthest to the right in the photo are directly connected to a DC solar powered pool pump, erasing the losses caused by inverting DC to AC. The DC pool pump is a most opportune use of direct connection, since algae growth is proportional to sun exposure; your pool pump moves the most water when it is needed the most.

It was not originally expected for that array to supply my entire electric budget…but WHY NOT? It was only after I installed that array that I started thinking about the concept that should be impressed upon every prospective PV buyer--every kWh conserved is one that you don't need from a solar array. Once you are in this mindset, a whirlwind takes over. I previously discussed a society mentally disconnected from its resource and waste streams. The freedom of our society means complete obliviousness to power consumption. We might have a "feeling" toward energy efficiency, but until we look at the actual numbers of our consumption, we will probably only be taking baby steps toward fixing our issues.

Hemorrhaging Power - While my first actions were qualitative, I would recommend folks to, as a first step, buy a cheap meter to measure power consumption, i.e., a 'Kill-a-watt'. You plug it into the wall and you plug your appliance into it. Its digital readout tells you how many watts of power the device consumes in all its various modes. Most devices suck up 'vampire' energy, even when they are turned "off". I have found that the "sleep" mode for our coffee pot and printer are very effective...the TV and DVR, not so much.

My focus on vampire loads intensified after I read an article about DVRs. DVRs aren't just vampires; they are trolls, spreading false information--and don't expect that anyone at the cable or satellite company will know about or advise you on this. It is false to think that your DVR is ever "off." Maybe they will get smarter someday, but since they are always waiting to instantly record a program, they are basically on perpetually. Long before we got the Volt, I did some calculations and found that putting my DVR on a kill switch would enable me to drive my Volt for free for TWO MONTHS of the year. WOW!! So, buy some power strips and/or uninterruptible power supplies (UPS). Get rid of excessive vampire drains by completely killing the power to your TV, stereo, etc.

As mentioned, I first started to kill electric loads out of principle. WHY NOT let the sun dry my clothes? After all, a Florida home with air conditioning doesn't need any more waste heat introduced inside the dwelling. Donna had long complained about the effect of the electric dryer on the longevity and wrinkling of certain clothing items and thus already had drying racks, as shown at right. The only time our electric clothes dryer is employed is when we can't wait for the weather to cooperate.

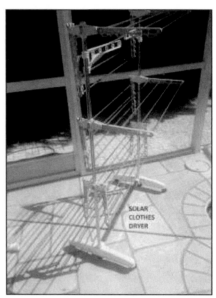

Figure 202 - Solar clothes dryer.

Since I was already getting free water from the sky, getting electricity-free HOT water made even more sense, so we purchased a solar hot water heating system. While we were initially happy with it, subsequent advancements in heat pump water heating have caused me to recommend going the latter route. Multiple failures of relief valves and pumps in the solar system have made it much more frustrating than expected. Instead, I would recommend getting a few more PV panels in the roof area you would have devoted to the solar hot water heating system. Another consideration: excess hot water cannot be sent to the grid like excess PV-generated electricity can.

Televisions and air conditioners are no-brainer upgrades when it comes to shopping your way to net zero. Cooking requires a bit more education. Over half the energy you use in heating your food is wasted, unless you use induction cooking. With induction cooking, magnetic interaction takes place to heat the pan instead of the cooktop, doubling the energy that goes to cooking the food. Of course, this does require the proper metallic cookware. But besides the benefit of energy savings, adjustments to temperature are nearly instantaneous, cooking is faster, and the safety and cleanliness of the cooktop is enhanced since it doesn't get near as hot and therefore doesn't burn spilled and splattered food as readily. For best results, we recommend you have a separate induction cooktop and convection oven. In our second home, we have a standing unit that combines both functions, which does not work as well.

Added Dimensions - And now that I have brought up that second home, let's briefly discuss how its presence affected our net zero considerations. Actions we took at that home will be discussed later, mostly following the chronology. For all intents and purposes, our net zero focus started with the Florida home, proceeded to the business, and concluded with the NY home.

Donna and I did things in reverse of what most snowbirds do. We are from up north originally, and it was my naval career that took us south. We enjoyed the south for most of our adult lives, as my first duty station was in Orlando, FL...followed by assignments in Charleston, SC, where we moved, in part, for Donna to attend dental school. For a number of reasons, following the dental path made more sense, and I left the military in 1994. Since we enjoyed Florida, we decided to move there for Donna to set up practice. The next dozen years saw the building of our house, the acquisition of a 32-foot cruising boat, and the growth--and ultimate departure--of our two children to move on with their own lives.

In the summer of 2005, we had chosen to take our timeshare week up north, and we started pining for the times of year we missed "back home": summer and fall. We analyzed our situation--with the very seasonal clientele we have in Florida--and realized that a second home made sense for us. Uniquely, it was that boat of ours that dictated the house search, as we wanted a location on the water where the boat could be driven from the dock in Florida to the dock in New York. Ironically, that boat never made that journey. It is now probably becoming apparent that when you start down a path toward electric drive and rejection of fossil fuels, you realize how much a boat that burns 30 gallons of gasoline an hour just doesn't fit anymore.

The logistics of our occupancy are such that I set my own schedule and can work from anywhere there is an Internet connection. I am in New York from mid-May until early December, leaving after I have finished deer hunting for the year. Donna takes multi-week vacations four times during the summer and fall, with her practice staffed by folks who consolidate the meager summer appointments into the weeks in between those vacations when she is back in Florida. Anyone who owns a lawn and a pool knows that I have just described a scenario that requires a paradigm shift if you don't want large bills from outside services to maintain multiple properties. We have no lawn service, pool service or snow removal service. The road to net zero is just as much about low maintenance as it is about anything else.

Lawn Gone - I love that phrase, and I hope you will embrace it. The comedian Gallagher would have had a field day with it. He would point out the absurdity of words that looked identical and sounded different. Should the above be "Lawn Gawn"? The absurdity of the language is akin to the absurdity of a home lawn. The labor and costs of trimming, fertilizing, and killing bugs for a plant you can't eat made no sense to me. There is no bigger non-agricultural source of pollution and water usage for plant life to be found in our society. And at its best, it only looks adequate when it is completely uniform. One small bug infestation in the middle of your lawn screams for attention. Seen from the street, it is a detriment to the appearance of your entire house. Bushes and flowers represent a combination of colors, shadows and depth to the eye. The same goes for the house trim. If your house and other landscaping had the same uniformity as your "flawless" grass, you would go out of your mind.

When we built our Florida home, there was a minimum turf requirement percentage dictated by code. Fortunately, drought and water restrictions killed that requirement years ago. I fought the turf battle for years, but it was a losing battle. The first thing to go was an area in the back yard at the intersection of two walls. The merciless sun bounced off those walls and kept killing the grass in

Figure 203 - Artificial turf eliminates pesticide and fertilizer runoff and requires no maintenance.

the area. So, I put in a hardscape patio there. Next came the constant battle with chinch bugs. In days, the lawn went from slight browning to several square yards of blight. Chemicals. Water. Time. Before we bought the New York home, I had ripped out more turf to put in paths, raised planting beds, and shrubbery that could withstand a dry spell. Drought-tolerant planting beds are the embodiment of net zero and the average American lawn is the utter contradiction of the concept of sustainability.

By the time we purchased the New York home, I had realized that the way to get rid of most of the rest of the lawn was by installing a circular driveway. The city inspector came to my door after the driveway installation, and I brought up my plans. (Note: Your road to net zero is NOT lined with bureaucrats with their hands up saluting you; it is as likely that their feet are out there to trip you up.) I explained what I was doing in the backyard, and she informed me of the requirement to have six feet of sod next to my seawall. I stated that, due to chemical runoff, I was forbidden to fertilize within ten feet of that seawall. All she could do was shrug her shoulders, because it was clear that different city ordinances were in conflict. She directed me to a city experiment using artificial turf, and that is ultimately the route I chose.

Currently available turfs have moved way beyond the sporting-field turfs of the past. The best products, like the "Foreverlawn" VR product we had installed, vary the coloration between grass blades and include a brownish thatch layer inside. It looks so real that it fools the chemical lawncare people. One asked from his truck, "How do you keep it so green?" We have about 1000 sq. ft. in the city's right-of-way next to the street and another 400 sq. ft., in two sections, in the back. As the accompanying photos show, it enhances the overall appearance of the property. It is expensive, but my maintenance can be as little as an annual sweep in that right-of-way.

Metallic Solutions - The services that cut your grass in Florida know a whole lot more about carburetors than they do about chinch bugs. The stress of dealing with a lawn does not end where the lawn mower blade hits the grass blade. Likewise, the services that deal with Florida pools know only one solution to water cleanliness: chlorine. We were fortunate that a family friend who worked for a fiberglass pool company educated us before our pool was installed. The impervious, chemically inert

…berglass make pool maintenance much easier than concrete pools, which require you to …ly battle alkalinity, as well periodically resurface the rough-textured material. Nonetheless, even …tic pool has issues with algae growth, and we did not desire our neighbors' solution to hire a …vice and keep the water so toxic that your eyes burned every time you went for a swim.

Until we bought our second home, I fought the conventional chlorine/algaecide battle with our Florida pool. I kept the right balance of chemicals to keep the pool in pretty good shape and almost always harmless to your eyes and swimsuits. But it was an ongoing effort, and acquisition of the NY property required research into something better. I have a relative with a salt generator setup, but it was never appealing…you now had salt water and chlorine--two things I didn't want in my mouth and eyes. The salt system was less work, but… Then I came across a copper ion system that seemed like the best solution. The accompanying photo shows the copper plates on top, which add ions inhospitable to algae growth. The bottom plates oxidize the water for clarity. At first, I tried to follow manufacturer instructions, which involved occasional algaecide, but later I adapted my own regimen that is much less work--a single jug (approximately monthly) of chlorine. My belief is that copper-resistant algae gain a foothold and the chlorine shock kills it and "burns" it up within hours. Liquid chlorine dissipates rapidly, and the pool water is odor and irritant free the next day.

Figure 204 – Copper ion treatment for pool.

Sustainably Comfortable - Climate control is an obvious contributor to one's carbon footprint. The typical Florida home has a heat pump for climate control, which is normally equipped with resistive heating elements for those few cold days. The electric heat pump, electric water heater, and electric cooking appliances add up to a lot of electricity demand! Adding sufficient solar panels can wipe out your footprint in short order, with a minimum of steps.

While many might think that two homes expand a footprint, they would be overlooking the role of climate control. When Donna is up in NY during summer for weeks at a time, the Florida air conditioning is turned off. Likewise, when the NY home is vacant for the coldest winter months, the pipes are emptied of water and ambient temperatures take over. The result is we watch the energy usage fall to what is required to run refrigeration to keep food preserved in the vacant house…and that's about it.

Traditionally, the higher latitude residents are the ones who have been forced to embrace the fossil fuel monster to keep warm. Furnaces have burned coal, oil, natural gas, and propane for decades. It should be kept in mind that while wood is a renewable resource, the residents of our planet have denuded forests since long before the industrial revolution began. We can only use this renewable resource for heating if we start thinking about sustainability. I'll get back to that.

Our NY home does not have a natural gas line and used propane at the time we purchased it. We soon discovered how cumbersome and expensive it is to have fossil fuel trucked in to keep us from freezing in the winter. That makes Florida a much better option for us in the winter.

Our experience illustrates that climate change is a fact. In the few decades we have lived in Florida, we have noted the air conditioning is used considerably longer into the winter season, and the heat pump is hardly ever used now to heat the house. In years past, we had to heat the house for a few days or even weeks, but those days are gone. The space heater that used to routinely sit in the master bathroom during the winter months hasn't come out of storage in years. It no longer gets cold enough, long enough, to drop the temperature of the house. The formally chilly days are now moderate, and the formally moderate days are now warm.

Meanwhile, our NY home security cameras show that snow is no longer a season-long occurrence. Decades ago, when snow blanketed the ground come December, that ground didn't reappear for months. Now, the Great Lakes stay unfrozen for much longer. The few feet of snow from the resulting blizzards is around for a much shorter time. The climate changes are unmistakable--and alarming.

The Eco Sensitive Dentist - If I could find an original 'green' moment at the dental office, it would be the day I got the electric bill after a cold snap in Southwest Florida. It was February 2008. It was the only time our bill ever went over $500, and I was at a loss to explain why. I realized it was time to educate the staff about what is known as the demand charge.

One chilly morning the staff put the heaters on for about 30 minutes. Business owners are billed for the amount of electricity they use and the RATE of that use. Specifically, whatever the meter shows as your highest usage for the month, even if it is only 30 minutes, becomes the demand charge, and every kWh used that month is charged at that highest rate. In our case, use of those heaters for 30 minutes resulted in a demand charge of nearly $400 for that month. 80% of that month's bill could be directly related to using those heaters for 30 minutes! Needless to say, the employees were banned from touching the thermostat after that. Security-coded thermostats now ensure no unauthorized adjustments.

Figure 205 - Solar installed at the dentist office.

Two years later, we installed a 13 kW array (57 panels) at the dental office—the largest array possible for our location. We expected the array to cover most of our office electrical needs. However, after all the subsequent changes we made, it turned out that this array size was close to optimal for us. To reemphasize a point made before, a kWh saved is better than a kWh that comes from solar. This is even more important at a business, due to those demand charges. This became painfully clear with our first electric bill after the array was installed. We actually had a surplus of electricity for that month, but the bill was still over $100, due to demand charges.

The next step for the office was to replace the conventional water heater with one that uses a heat pump. The heater can be set so that the system will operate without its coils, ensuring it never draws more than 550 watts--the electricity used by the heat pump alone. This significantly lowers the electricity demand and thus our demand charge, as the resistance coils use 10X more power.

Next, we tackled climate control. When the office was under construction, the contractor determined that two 4-ton AC units were required. However, my experience at home showed we didn't really need both units, since my home interior was comparable in space and one 5-ton unit worked just fine. When I presented this concept to the contractor, he refused to make the change. He had his cookbook approach to commercial spaces, and he was not going to budge. Since I was the customer, he was either going to be dismissed or do it my way. Ultimately, we reworked some ducting and the 5-ton unit has done just fine for the 3000 sq. ft. it cools. When the new system was installed, we disconnected the heating strips, preventing the demand rate debacle previously described.

One of the lessons here is to remember that old schools of thought die hard deaths. Each change in methodology requires a revisiting of the cookbook. For example, a person at rest gives off about the same amount of heat as one incandescent light bulb. Replacing every incandescent light bulb in the office was going to eliminate a lot air conditioning demand. The air conditioning contractor just could not get that through his head. At that time, we installed CFL bulbs, but even cooler running LED bulbs are now prevalent throughout the office.

Lighting changes helped lower the heat load in the office, but we wanted to do more. Therefore, we sprayed foam insulation onto the underside of the roof decking and premium, seemingly transparent solar film was applied to the windows. At first, we chose to save some money and only applied the film in the sunnier parts of the office, but it quickly became clear that lots of solar energy was entering even the non-sunny side of the office, so we added film to those windows, too.

Anyone who has paid a visit to the dental office is aware of the obnoxious light that is shined directly into their face, so the dentist can perform his/her work. Around the time of our office makeover, LED lamps were starting to appear. While an easy decision for those starting from scratch, it was a much tougher choice for us to discard working lamps just to improve energy efficiency. When I asked Donna what she wanted to do about the conversion, she almost irritatingly replied that she wished it was only about energy efficiency. It was a huge expense, and if the only thing to be gained was reduced consumption, she would have preferred to skip it. However, LED lamps offered many advantages--again a significant

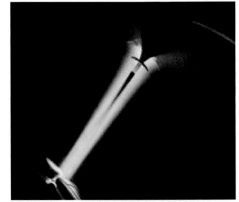

Figure 206 - Focused LED Dentist lamp.

revisiting of that cookbook. Since LED lights are much more focused and directional, the patient experience is drastically improved. The mouth is now the only thing illuminated, and there is no spillover light blinding the patient's eyes, as shown. Also, LED lights range a color spectrum from cool (bluer) to warm (yellower). Since ultraviolet light is used to cure many dental materials, the dentist can now choose LED settings away from that spectrum and provide him/herself more time to work with the material before it sets up. Finally, the old lights were hot--very hot. If the dentist or assistant missed the handle and touched the housing of the old lights, they easily got burned.

About a year after the above alterations were completed, a ballast (transformer) was failing in one of the fluorescent fixtures, and it dawned on me to check into the current status of LED tube replacements. I realized it was a no-brainer to simply snip the ballast wires, bypassing them, and installing LED tubes. The cost of the tubes was offset by the cost of the ballast. Since then, of course, all LED options have gotten much cheaper, and it is crazy to even think about buying fluorescent bulbs, given the disposal headaches that these mercury-laden bulbs involve, the audible buzzing from the fixtures, and the annoying flickering that inevitably appears near bulb death. I ended up replacing several sockets for the tubes, but that headache is gone now. One just needs to buy the right LED tube replacements.

I decided to swap out the fluorescents for LED tubes all at once. I did this in part because of that demand limit issue. The magic number in Florida is 21 kW of demand. If you permanently maintain less demand than this, you can get yourself shifted to a different rate system where the only thing that matters is kilowatt-hours--the amount consumed. My prior modifications had gotten me down to 10-12 kW of demand. However, I was thinking about adding a car-charging station and that 7.5 kW addition would get too close to making us a demand customer again. It turned out that most fixtures could get by with half the tubes, due to the increased brightness of individual tubes, and each LED tube was half the power consumption of its fluorescent equivalent. I was now using just 25% of the energy that I was previously using for lighting, and my demand number was down to 8-10 kW. The last, and perhaps best thing to mention, is the longevity of LEDs. I expect most of this lighting will last for as long as we own the business. The road to net zero is actually a relaxing drive, with many sustainable options requiring less effort compared to the old way of doing business.

The Net (Zero) Effect in Florida - Both the Florida home and the dental office have solar PV arrays connected to the grid via net metering. We use no other power or fuel of any kind, not even for yard equipment. We do have a yard service that prunes our trees and bushes a couple times a year, so

they probably use a couple gallons of gas maintaining the house. All commuting is electrically powered by the home array, so examination of our electric bills and meters gives a thorough analysis of our entire Florida carbon impact-just plain negligible for home, and next to

Figure 207 - The SW-facing roof is the array we installed after calculating what we need to drive.

negligible for the office. In years past, November through May saw both home and office as net suppliers of electricity, with the remaining months making us net users of electricity. The overall effect is determined in December when the account is zeroed out if we have a surplus, which is paid back to us at wholesale pricing.

This brings us back to how rapidly we feel that climate is changing. Until a couple years ago, we could expect a few dollars' worth of credit from that home's December bill, with the office credit being a bit less or slightly in the utility's favor. But we now must air condition at times that we did not before. Perhaps it is because our utility energy consumption is near negligible that we noticed the transition from being slightly net suppliers of grid electricity to slightly net consumers of it. The rapidity of this transition is alarming.

The World is Fracked Up - I have left the ideology of becoming green until the last section, having pragmatically approached net zero, but this is where I firmly shift course. It is hard to pinpoint the conversion to becoming the zealots that Donna and I are at this point, but the change was well underway by the time of the Deepwater Horizon tragedy of 2010. For months, we watched the horrible, slow-motion toxic events play out and made the incorrect assumption that the world was finally about to dump its fossil fuel addiction. Surely, as the pristine Gulf beaches were coated with goo, people would change. Surely, as hundreds of square miles showed visible oil spill, people would change. Surely, with all the dead animals, people would change. Surely, with all the lost revenue to the Gulf States, people would change. Those changes didn't happen.

At the five-year anniversary of the disaster, a wonderful documentary, "The Great Invisible," was made about it. Here is a quote:

> "Generally, it takes some kind of a traumatic event to change people's behavior. I had hoped that the Deepwater Horizon was going to raise everyone's consciousness. But, it didn't. We had a moment in time where everybody was paying very close attention to where we could actually change the way we think about burning hydrocarbons in this country. And the political pressures, from the Congress, to the Senate, and in the White House, pushed those issues to the side and voted to stay the status quo."

> Bob Cavnar
> Oil Executive
> Author, Disaster on the Horizon

This is the sentiment I had. Previous to Deepwater Horizon, we were becoming intensely aware of how addicted we were to the fossil fuels that were killing us and the planet. After Deepwater Horizon, we saw fossil fuels as toxic poisons that needed to be discarded as soon as possible. I wish the world would have seen the reality, but toxic fracking made oil cheap and everyone forgot about Deepwater Horizon.

The California drought came, and no one cared that farmers were outbid for water by frackers who needed millions of gallons of water for each of the thousands of holes in the ground. Because gasoline was cheap, no one cared that well water was becoming undrinkable in fracked areas. News viewers became transfixed with the dramatic acts of terrorists who would kill dozens, while relatively boring climate news told us that billions were in peril if we didn't break our fossil fuel addiction. People proved to be anything but strategic planners, easily distracted by the dramatic instead of the important. And when the drama wore off, memories proved short.

And somewhere around this time, Donna and I became much more aware of the karma concept: what matters in our existence is what we have control over and how we respond. We have to make the right choices for ourselves, regardless of whether the rest of the world gets it or not. So, fossil fuels would be directed out of our lives to the fullest extent possible.

Before I get back to the actions toward net zero, I need to make one more comment on Deepwater Horizon that I alluded to before. Every night for months, the news showed the horrible mess in the Gulf. Those cameras visited the beaches and the marinas, but they seemed to never leave water's edge by much. And they didn't venture down the rest of the Florida peninsula. If they had, the viewer would have seen empty places of business that had nothing to do with tourism. The timing of the spill could not have been worse, coming on the heels of The Great Recession. It wasn't just tourism that suffered. People just weren't here in Florida, conducting business. The lowest of our lows came well after a recovery started elsewhere. That oil spill cut our income down to half of what we were making before the recession. Oil is toxic on so many levels.

Fortunately, I had the numbers to prove the loss of revenue, and we secured a settlement, which I considered "blood money". It only seemed right that the majority of the purchase price of our Tesla Model X was paid for by British Petroleum. We had been fine with the two 2012 Chevy Volts we had, but the settlement allowed them to go and save gas for adult children, and we got rid of almost all the gasoline we had left in our lives when the Tesla arrived. Our current vehicle inventory now includes an updated 2017 Chevy Volt...and that 2008 Tahoe Hybrid is still around, but barely driven.

The Great White North - When we bought the NY house, I started with a goal of making that home energy neutral, as we had done with the Florida home and business. While the New York boat dock never saw that cruising boat I mentioned previously, the dock we extended, for it seemed like a great place for a solar array. The lot itself was covered by ancient, expansive oak trees. The only option for solar was out over the sunny water, away from the trees.

With the solar goal in mind, I learned that subsides were based on historical electric bills. I could not embark on a solar path until the house had been entirely shifted to electricity and off propane. Digging up of the front yard to rid us of septic was the catalyst for the next several steps. We had the guy who dug our sewer connection also pull out the large propane tank buried in the front yard. The grass went away and, as we did in Florida, we installed a circular driveway, ridding us of grass out front. This was much more necessary in NY because the house came with no garage, and parking was an issue.

The house had a conventional fireplace. We knew that the open flue made a conventional fireplace a net loser of energy. Enter a secondary combustion fireplace insert.

The secondary combustion fireplace insert utilizes this normally wasted gas, as well as the rest of the wood, so that the wood burns more completely, with less wood and producing much less ash and smoke. Secondary combustion, or gasification, is the only sustainable and neighbor-friendly way to think of wood as a heat source. Even if we replant enough trees, the smoke from conventional fireplaces would make the air impossible to breathe. If you stand outside the fireplace at our house, the emissions from the chimney are barely discernable in either odor or visual clarity. A few logs in the fireplace before you go to bed heats the home through the night.

Figure 208 – Secondary combustion fireplace insert.

Safety is not an issue, as the combustion chamber is sealed and there will be no embers flying out of the fireplace. Gasification fireplace inserts are not about the ambience; in fact, those pretty flames are a sign of inefficiency. Peering through the viewing window of our fireplace shows a few flames, but lots of glowing embers. The accompanying photo, grabbed from the manufacturer's website, makes for a good show, but is not an accurate representation of what you would normally see. If you do want the show, instead of the heat, you would just leave the door open for lots of picturesque, if inefficient, flames. The fireplace does come with a screen that fits over this opening for such occasions.

The bulk of heating of the NY house was previously from a central propane furnace. It still sits in the basement, disconnected, available for a future homeowner wanting a large propane bill. Keeping in mind that we are not ordinarily in the home from January through April, the largest part of our heating is now done by two ductless, mini-split heat pumps. Ordinary heat pumps just don't cut it when the temperatures dip down into the teens. These mini-split heat pumps provide hot air down to even frigid temperatures. However, if we are at the NY house during the coldest winter months, we would be running that fireplace insert, as well. Finally, we also installed a conventional, ducted heat pump for the upstairs for occasional visitors who might show up. [The master bedroom is on the first floor.] This system also provides a cross-connect to intake air from the normally warmer first floor.

I earlier mentioned the combination convection oven/induction cooktop that we purchased to replace the electric oven. They had a propane clothes dryer, but we purchased an electric one--rarely used because most clothes drying is done outside, unless the weather precludes it. The last appliance to go was the propane water heater, replaced by, of course, a heat pump water heater.

There were never any gas-powered tools purchased for the NY house. The lawn mower was a Mark-powered reel mower. With the circular driveway, the front lawn disappeared and the last of the grass disappeared entirely in 2017 when the backyard was redone with stone terracing, shrubbery, and a few patches of artificial turf, like the Florida home.

The Tesla is driven to NY and acts as the primary vehicle. While the Tahoe Hybrid stays in NY, it is a backup only. Donna flies in and out of Buffalo, 120 miles away, where her family lives, and it is occasionally more convenient for her to drive the Tahoe there and leave it for her return trip some weeks later. It is used so little

Figure 209 - Heat pump water heater.

that we use it on purpose to prevent the gas from getting stale. We now burn less than one tankful per year. For the first four years of Volt ownership, and during the first year of Tesla ownership, home electricity provided much of the power for transportation. However, Tesla's continued expansion of its Supercharger network now means that there is Supercharging for most trips. I drive to accommodate my cycling hobby of finding rail trails, so the Tesla gets much more use than one would probably expect in a non-commuting household.

The propane is gone, and I use approximately 35 gallons of gas annually to feed the Tahoe and Volt. All other fuel is via the electric utility. My power bills show approximately 150 kWh per month of use during the dormant months, approximately 400 kWh once the house again becomes occupied, and a bit over 1000 kWh once the heat pumps are involved, with approximately 20% of that figure devoted to driving to go deer hunting. It is difficult to evaluate the before and after energy picture. However, I do remember my higher propane bills being over $400. My current highest electric bills, which include transportation, do not rise over $200.

While our large boat was sold before ever reaching the NY home, I did keep the dinghy. Initially, I fought with the 6-HP gasoline motor until I had had it with pulling and pulling to start the motor. I could devote a chapter just to this item, but I'll try to be brief. I was terrified each time I pulled that heavy, bulky outboard from its storage to attach to the dinghy. I always worried it would plunge into the water as I maneuvered it on the slippery, bouncing swim platform. A few years ago, after it failed to start, I looked up electric options. I found

Figure 210 - Dingy with outboard electric motor and extra battery.

Torqeedo. The travel motor I got incorporates a lithium battery in the housing which probably gets me about 15 miles. I also bought an additional battery. It is not as fast as the gasoline motor, but it is fast enough to kick up a wake. It is no trolling motor. It separates into 3 pieces, which are very light. If I still had the cruiser, I would no longer worry and struggle. This electric motor is a no-brainer for use on a tender, where the average trip is normally in and around the marina. My all-day trips to the nearby state park cost me about a dime's worth of electricity.

My journey to net zero is about as far as I can go. All the lights are LED. No burning fossil fuels for warmth. Electric transportation. There are a few items that will inevitably take another step, like that backup Tahoe being replaced at some point.

There is one issue that made things difficult. During the few years I spent electrifying the NY house, I was stopped dead in my tracks by bureaucrats at both the state and federal levels. Market solutions and peer pressure will always be preferable to regulations.

The Bureaucrats - With bright expectations (no pun intended), I contacted a central NY solar contractor. They had to do some significant work to figure out how my system could be set up. It was going to be a post mount on each end with support along the horizontal axis of the array. I worked with the designers so that the bottom of the array would be about 3 feet above the dock hanging out over the water, and the top of the array would be 10 feet above the dock, angled back to be above the head as you walked on the dock. It would have 28 panels and produce about 5 kW, like our initial Florida array. While it would be more expensive than I expected, I understood this

Figure 211 - The dock area would be perfect for solar.

array would be more complicated than a roof-mounted array. But this was no longer about the money. I was trying to blaze a trail...and boy, I did not know how far off the beaten path I was about to go.

Figure 212 – The state wanted me to cut the 300-year-old oak tree for my solar installation.

While the dock may belong to the homeowner, the water underneath does not, and so began the process of applying to three different state agencies, as well as the US Army Corps of Engineers in Buffalo. Because I had lengthened the dock the year before, I was somewhat familiar with the people and the process. I had early expectations of success from the COE fellow, who would release the actual permit. He was ready to approve it because the dock function would not change. It was still a dock.

Then the state got involved. The irony, and lunacy, of the situation is that I could erect a solar array and generate all the power I wanted if it was used on the dock. That's right: you can pull up your yacht, use solar panels to provide shore power to it, and everyone was fine with it. However, if you ran the power line from that array to your home, you were screwed. Two of the three state agencies did not have a problem. However, the NY Department of State stopped me from erecting a solar array by referencing a federal Coastal Management document. In essence, they told me to cut down 300-year-old oak trees to keep that array off the dock. I examined their guidelines, and there were sections about scenic beauty and preservation, which I tried to reference in appealing to the insanity of their judgment, but nothing worked. The avenue of appeal at the federal level was, of all agencies, NOAA. I thought this should be good for me, since this is the one agency that seemed to know about climate change and care about replacing fossil fuels with renewable energy. I used the language in their own document...but again, nothing worked. They said they didn't want to overrule the state...and the state told me they were siding with guidelines from the federal government. No one wanted to take responsibility in this circular, fingers-always-pointing-to-someone-else mess. No one would put it in print that I needed to cut down trees, but it was the "elephant in the room" that everyone ignored.

While I expect I could have won approval in court to erect my array, the array itself was already going to be quite expensive. From a financial standpoint, the array on the dock was probably not going to pay for itself. Adding legal expenses would drive the cost to unreasonable levels. While I did contact various consumer advocates, news agencies, politicians, renewable energy foundations, and even a college professor who made the news by advocating for old trees, nothing changed. Ultimately, I elected not to chop down regal, old trees, and I found a utility that sells electricity produced from wind, hydro, and landfill gas. So, my approximate usage of 6000 kWh per year--which includes all home and transportation power--is still close to carbon free for the NY home.

Regulations are not the cure-all that some believe. Regulations are the playground of lawyers, condemning all results to time delays, while courts rule on lawsuits by both sides. Likewise, free markets will never solve problems when no value is placed on human health and suffering or on the environment. As with so many things in life, it falls to the moderating voices to find the best solutions. I consider myself one of those voices. I like to refer to myself as a "Radical Moderate."

The Radical Moderate - I use the term radical because so many at the poles of reason assume that those of us in the middle are not passionate about our actions and ideals. They assume that a position that is altered is one that is not committed to, mistaking that we Radical Moderates actually LISTEN to conflicting or new information. The reasonable person will change their mind when circumstances warrant. The polarized, tribal folks trust us folks in the middle even less than their polar opposites. At least the folks on the other end of the spectrum are predictable. We would be unreliable turncoats, ready to shift our position if a new study points out the folly in an old viewpoint. Being a Radical Moderate means there is no tribe to accompany one on the journey through life. You are alone, except for those around you who listen to their moral compass and treat people like they wish to be treated. The Radical Moderate, the ultimate pragmatist is, concurrently, both the most agreeable and the least agreeable person you will find. We are the most agreeable because we will try to find some common ground and forge solutions that benefit everyone. We are the least agreeable, because we can't be expected to agree with others all the time on anything. Worse yet, we might withdraw our agreement if the data changes or our moral compass tells us to.

Polar, tribal people will have no end of forgiveness for each other, within the group, because the common goals of the group overwhelm flaws, lies, backstabbing, cheating, etc. In fact, the tribe will move together so well as a unit, they won't even realize when they flipped positions from previously firmly held beliefs. Polarized folks remain so firm as to embrace conspiracy theories rather than consider data that is contrary to the group-think. But participation in group-think does not allow for differing opinions on individual issues. Self-identified liberals place climate change high on their list of concerns, but budget deficits are not even thought about. Self-identified conservatives flip these two issues completely.

As the Radical Moderate, I would point out that BOTH groups will sacrifice their own progeny. One group will leave them with an economy in tatters, and the other will leave them with an environment in tatters. Only we in the middle have any hope of getting these two poles to stop shouting and start fixing. As such, I will end with a positive story to offset the person who screamed at me as described at the beginning of this chapter.

Think about an all-electric Tesla or extended-range electric Chevy Volt going wherever its driver finds fun and recreation. You might picture hiking or biking. Now, think about a harvested deer being carted off by a hunter on his way home from the woods. You probably picture a pick-up truck as the vehicle doing the hauling. That fellow who screamed at me thought he had the world all figured out and properly boxed up. There are red necks and there are tree huggers and the lines are clearly drawn. Because I drive a Tesla, I am already a completely known commodity to his polarized view.

Now merge the two mental images from above. It was me driving that Volt with a deer on the back until it was me driving that Tesla with a deer on the back--and I loved watching people's heads explode as I blew up their paradigms. I watched a lady passenger in a car behind me snap a picture of the deer on the Volt. I had a bearded guy in a pick-up truck speed up, slow down, pull alongside and give me a gleeful smile and a thumbs up. At the deer processing place, a guy came in and

Figure 213 - An environmental minded deer hunter.

waited patiently while I spoke with the clerk. When I was done, he smiled and asked if that was my Tesla outside. When I said yes, he said, 'I just want to shake your hand.' It was clear I had just busted his paradigm, to a very delightful ending. There are people out there who want to see the world change, but maybe just haven't seen a picture where the capitalist and the environmentalist are both right--and both wrong. I hear Mother Earth calling...and her patience is running out. All sides need to sit down and talk before the table, and the entire room, go up in flames.

What goes around, comes around - I thought I had wrapped up my discussion and my thoughts, until the ouroboros came to mind after a visit from an insightful solar technician the day after I finished typing the preceding paragraphs. Donna is intrigued by mystic imagery, and she likes the symbolism of the snake eating its own tale: the ouroboros. This is the symbol of infinity, of no ending, of returning to the beginning, of re-examination. It is just so appropriate that a solar tech visit in 2018 brings me back to expectations and motivations from a quarter century earlier.

After concluding the journey down a road where politics and passions have come to the forefront, it is worthwhile to reiterate that my road to net zero started with my earliest frustrations when confronted with waste. From my childhood, I would always think of the waste of running the hot water in the faucet until all the cold water made it through the pipes and hot water actually came out the spigot. When it came to designing and building our Florida house, it struck me that the master bath was a long way from the hot water heater in the garage. I brought this up to the builder, and he suggested installing a recirculation line with a pump that would bring hot water to the bathroom. Great. "Add it," I said.

After a dozen years, the recirculation system became unusable when we had the solar hot water heating system installed. I brought this up to the solar contractor, and he replaced the pump, but that did not solve the problem. I was pretty sure their solar hookup was introducing air into the system and the pump was air-bound. But no one could, or would, investigate further, and I just disconnected the pump.

But I still hated wasting all that water. Therefore, we kept a five-gallon bucket in the bathroom. The hand-held shower head was put in the bucket, and the water was run until actual hot water was entering the bucket. It was a predictable amount of about 2-3 gallons. That water was then used for toilet flushing, so nothing went to waste. Note that we only used this bucket during the dry season—November to June, because there is lots of excess from the cisterns during the wet season.

The solar contractor returned with the expectation of again having to replace the solar pump. I had told the solar company repeatedly that I thought air was binding up that pump, but no one put their mind to the problem--until a most insightful tech showed up. I took him over and showed him the small leak on an exterior wall where the solar company had disconnected a heat exchanger. That heat exchanger used waste heat from the air conditioner to heat the water. I was pretty sure they needed to deal with the capped-off connections. Ultimately, some valves were turned, isolating that severed connection...and the introduction of air into the system was stopped. While we had not been discussing the recirculation system at all, I was suddenly reminded of my suspicions that the system had become air-bound. After flipping a switch and realizing that the recirculation system was again working after years of being idle, Donna's joy came from realizing she was done dealing with the bucket. Yeah!

I mentioned flipping a switch...and this is where the ouroboros really comes in. I hadn't thought about it in years, but that is the switch that I personally installed. It is a switch that represents all my discussion here about self-examination, re-examination and striving for constant improvement. It is a switch that proved how un-insightful you should expect others to be. It is a device symbolic of how, ultimately, I had been burned by lots of 'experts'--from certified financial planners to that air conditioning contractor at the dental office. I had a phrase that I used to use when trying to sell my management software to people who simply refused to use their own brains in their own best interests. "Success begins when you refuse to let OTHER people do YOUR thinking for YOU!"

Sure, our home contractor gave me a recirculation line to save water waste...but he gave me no way to turn off the recirculating pump. It took me weeks of home ownership before I realized that he had provided an always-flowing supply of hot water through an uninsulated pipe in the attic. Saving a few gallons of water had resulted in wasting countless kilowatt-hours of heated water instead. I really could not believe the complete lack of awareness of waste that this represented. My first attempt to solve the problem was a battery-operated wireless switch from the bathroom to the outlet in the garage. However, hitting the button provided no feedback as to whether the signal was received in the garage-- and it often wasn't. After a couple years, I got up into the attic, found the wiring to the outlet in the garage, inserted some wiring back to the bathroom and installed a switch that is now working. Putting my thoughts to keyboard made me realize how many of my own conservation habits have become second nature, as well as how few people think about their impact on the world around them.

The symbol of the ouroboros is how I would like the reader to remember my discussion. It represents the circle of life. My efforts are in large part dictated by the fact that there is no "my fair share" for the world and its resources. We don't inherit the earth from our parents; we borrow it from our children. We are not allowed to use it up to satisfy our selfish desires. That is what karma is all about: what goes around comes around, as the ouroboros' head comes back to its tail. We must re-examine and re-visit how we impact our world and our neighbors. We must never be complacent--fat, dumb and happy--but instead strive for improvement in all ways...always.

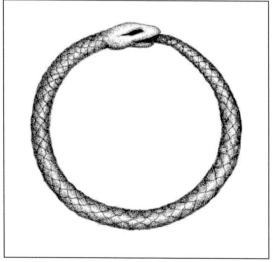

Figure 214 - The ouroboros represents the circle of life.

Chapter 14 – Do No Harm

by Sean Scherer

When we decide to do something big with our lives and pour our intentions and efforts into achieving that big thing … it often takes us on a journey we don't expect. It may not be predictable and may be fraught with challenges. When it's over, we may find ourselves standing on the shoulders of giants and enjoying the company.

Figure 215 - Queenie and me with her prior Prius in the background.

It's easy to get swallowed in tradition and continue doing things the way we've always done them. The world is set up to make it easy for us to do this, to avoid deviations from the norm. In fact, I've found that in some cases, it costs me more money, more time, and more effort to have a positive impact. It is way too easy to have a negative impact; the free-market economy has fostered this throughout the industrial era, and it continues today. Going to McDonald's for a meal is cheap, quick, and easy. Finding an old SUV to haul yourself, your family and all your things is simple; they are a dime a dozen, and gas has always been cheap in this country (relative to many other first-world nations).

When I was first asked about contributing to this book, I had a very specific idea of what it meant to drive to a "net zero" carbon emissions lifestyle. That idea was shaped by my passion for electric vehicles and the fact that I've been a long-time activist for and contributor to driving electric. I started my own journey when I fell in love with this girl and her hybrid and ended up committing to both! This was more than a decade ago, before any mass-market electric vehicles were for sale. Then we met a neighbor who worked on electric vehicles, and I have been "drinking the Kool-Aid" ever since–having personally owned and/or restored more than a dozen electric vehicles.

However, as I started spending the time thinking about this book, I realized driving to net zero is about much more than just the electrification of our automobiles, so very much more. The term 'driving' may evoke the automobile, but leading a "net-zero" lifestyle is a much more involved matter, with many issues one may overlook when focused on automobiles alone.

Robert Frost's famous words come to mind: "Two roads diverged in a wood, and I— I took the one less traveled by, and that has made all the difference." When I was a child, we read that poem in grade-school, and I really identified with the concept of taking the less traveled road - the one not tattered from wear, but the one with mystery and potential reward, the untrodden path! As a young adult and having just met my wife, Queenie, I had a chance to start heading down this road in the most meaningful and rewarding string of decisions I've ever made--decisions which make me swell with pride, decisions for which I have no regret. I intend to give an account of my story: how I went from your typical Wisconsin-born youngster to a caring millennial with a penchant for sustainability, whose greatest accomplishments in life include driving nearly exclusively on electricity for the past 10 years, consuming a plant-based diet, running a home on solar electricity, and being involved in initiatives for change at all levels. If there is one take-away I want to leave you with, it is *ahimsa*, a Buddhist virtue to do no harm that Queenie and I incorporate into our decisions and lifestyle.

The ahimsa-net-zero lifestyle is a practice that anyone can follow in order to cause the least amount of harm, no matter your situation. I hope this chapter inspires you to think beyond cars and try some of our ideas and tips toward a lifestyle that benefits everyone. Full disclosure: we don't commute by bike, live in an off-grid tiny house, grow our own food, and practice yoga every day (yet)! We have very different backgrounds, goals, and habits, but in all we do, we try to pick the path of least harm. You are doing something important simply by keeping an open mind while reading this book!

We believe the concept of ahimsa--to do no harm--is, in short, the net-zero lifestyle and impacts the consumption of water, energy, food, and other "stuff". It promotes activism and volunteerism and impacts our individual lives, as well as the entire community.

Quick Facts:
>   **Net Energy Emissions** – 11,000 lbs. - 75% less than average
>   **Home** – 3,000 sq. ft. in the Phoenix area of Arizona, USA
>   **Technologies Employed** –EVs, solar PV, vegan diet
>   **Annual Net Energy Costs** – $1900. Note: Just moved to a new home and solar is not yet on-line.

Our story - It was 11 years ago when I had my first glimpse of how the everyday choices we make can have major impacts on the world. It started when I met Queenie, an environmental engineer whose passion for preservation and sustainability was so brilliant, it ended my desire to carry on living the way I had for the previous 21 years of my existence. I met her when she decided to volunteer in Louisiana after Hurricane Katrina, where I was doing animal relief work. We have been together ever since.

Queenie first gave me the tools to care. She established the concepts of sustainability and how the choices we make in life have everlasting impacts on the planet, like a single rain drop leaving ripples on the surface of a perfectly still lake.

Queenie's enthusiasm permeated her everyday life. At first, I thought she wasn't real. Would a person really bring their own bags to a shopping market? Seek out locally-sourced foods? Collect greywater for reuse? Eat solely from the base of the food-chain? She was dedicated to the cause and would tell me, "It's just the right thing to do to cause the least harm (ahimsa), waste less, and vote with your dollars." How could I disagree with this fascinating woman? She intrigued and challenged me to examine the daily role I play in the planet. Do I want to leave this planet in worse conditions because I was here? Is it possible to exist with a small footprint? I realized quickly that I wanted to have a positive impact, even if ever so slightly. I wanted to leave the planet in better condition than I found it when I was born. This decision has driven many of the choices I've made through the years and has been a defining mantra of my life and outlook on the Earth.

At first, Queenie and I did not give much thought to Mother Earth until we met others who inspired us to find better alternatives. Our parents came from poor families, and none of them went to college, but luckily, they encouraged us to do so. I was one of five kids, and my family moved around the US a lot as I was growing up. Queenie was an only child and immigrated to the US after elementary school. We grew up following cultural norms, and any conservation we did do was to save money.

Queenie picked engineering after high school because she likes solving math problems. When she got to college, she realized she didn't want to spend her life creating more stuff like gadgets and chemicals. She transferred into civil and environmental engineering, where she learned about big environmental problems that take a village to solve. Her activism in bettering our environment developed professionally: She is a Leadership in Energy and Environmental Design LEED Accredited Professional, a Certified Sustainable Building Advisor, an American Rainwater Catchment System Association Accredited Professional, and a licensed professional engineer. She landed a career in energy and water conservation, specializing in performance contracting where it makes the most dollar, cents, and sense for campuses with huge consumptions such as military bases, airports, hospitals, prisons, and universities.

Saving water - Queenie's parents grew up when water was rationed in Hong Kong. Each household would fill up their buckets and carry them back for daily use. Fast forward to present, where this is still a daily practice for many women in developing countries. Women walk for miles on foot, transporting water on their heads instead of going to school. Recent predictions indicate that we will be fighting over water in the next world war because water is life. Recent news documents that large cities around the world are in a water shortage crisis.

In the US, with our mega-farms, lush green lawns, swimming pools, and drinking water delivered to the taps unrestricted, it is easy to under-appreciate water, even when we live in the desert. Water is relatively cheap in the US, largely because the Western philosophy believes water is a right, so water has been subsidized by our government, and its cost has been artificially kept low. In many developed countries, water is not potable from the tap because it is too wasteful to treat water to potable standards. Very few of us even drink our 11 glasses of water per day, so treating only the volume of water we consume internally makes good sense.

A great way to conserve water and save money is to reduce water flow in our faucets, toilets, showers, washing machines, dishwashers, and irrigation systems. Many municipalities offer rebates for investing in these one-time purchases and installations and for even advanced systems like rainwater harvesting, greywater reuse, low-water landscaping, and artificial turf. Even if you can't afford these upgrades yet, you can be more efficient in your use. Skipping the manual pre-rinse of dishes, loading your washing machines and dishwashers to capacity, adjusting your irrigation system to make sure you are not watering at the hottest time of the day or watering the sidewalk, and finding and repairing leaks are all great, low-cost options. A weekend project could be to channel rainwater to your trees and gardens by digging paths and ponds through the yard or building an outdoor shower and rerouting your shower and laundry water (with biodegradable, phosphate-free soap and detergent) to your trees and gardens.

Smart showering can save water and money. If you turn your shower on hot and go do something else for a few minutes, you are wasting gallons of water, including some hot water from your water heater!

If you flush your toilet during the bleeding of the cold shower water, it actually shortens the time for the hot water to arrive, because the cold water has been reallocated. Queenie collects the initial cold shower water to refill water bowls for our animals, water her plants, and for future toilet flushing.

These are just a few of the ways you can save precious water and money at the same time.

Transportation - I had always dreamt of being a doctor and was in medical school at the time we first met our neighbor, Greg, shortly after we moved to Phoenix. That meeting opened up a whole new, wonderful world for me. The evening we met Greg he showed me what was in his garage: a 1980's VW Scirocco converted to full electric by engineering students at Arizona State University and a GMC "G-Van" built by General Motors as a fully electrified passenger van back in the early 90's. Even though I have been passionate about electronics and technology for as long as I can remember and a passionate car lover and proud hybrid vehicle owner, I had no idea such vehicles existed. Learning that you could power a vehicle on electricity alone was a revelation, and my life was never the same from there forward. Within months of meeting Greg and spending time learning more about his vehicles, I was heading out to buy my first 100% electric vehicle, a 1998 Toyota Corolla that was converted by an engineer as his personal vehicle. It had a 20-30 mile range, and getting it home meant that I had to drive halfway back, park it at a friend's house to recharge overnight and then come back the next morning and drive it the rest of the way home.

Ever since meeting Greg, instead of studying as much as I probably should have, I would spend a lot of time tinkering on my latest home-built electric car project. Note that this was at a time before automakers launched electric vehicles to the masses. I did eventually graduate from medical school with a huge debt and started a career in naturopathic medicine around the time Nissan and Tesla debuted their first electric vehicles. I realized I could make a bigger difference in the world by pursuing a career in electrifying transportation. However, without an engineering or mechanical degree, my self-taught skills were a tough sell to employers. Queenie was under a lot of stress, as she had been our primary breadwinner for over 10 years, so she suggested I post on Craigslist that I could help others restore electric vehicles. That's literally how I landed my first break, starting up an electric car division at a local Uninterrupted Power Supply company.

From that point I went through over a dozen electric vehicles. I had to experience them all; I loved them all! At this time no manufacturer was building and marketing electric vehicles--there was no TESLA, no Nissan LEAF, no plug-in hybrids. When I told someone that I drove an electric vehicle, they were stunned, as they had never heard of such a thing and couldn't conceive how that could work.

I spent countless hours building, repairing, and improving my electric vehicles. I would buy one when it was no longer functional and make it work again, then enjoy it for a while and sell it for what I put had into it—or more. It was an enjoyable labor of love, but at the same time I always wished there were "better" electric vehicles out there, places to plug them in, ranges that exceeded 50 miles in a day, batteries that didn't melt down their terminals, etc. The lead-acid batteries of the past had more than their share of headaches.

When Tesla first announced the original Roadster and targeted a price under $100,000, I had a new dream-car. Bye, bye, Ferrari! I promised myself I would have one someday, and I eventually did. My first TESLA was VIN 531 and it was a Red Roadster 2.0 that had gotten into an accident with its front-end torn apart. I rebuilt the front end and enjoyed that vehicle (with a mis-matched blue bumper) for more than a

Figure 216 - My first Tesla and real EV that could replace a gasoline car.

year. I later sold it to an engineer at Apple in Silicon Valley.

Figure 217 – Working on a Roadster while at Gruber Power, where I restored and repaired EV's.

Queenie and I now drive a pair of electric vehicles. She drives the Nissan LEAF, and I drive a "ludicrous mode" TESLA Model S. We are living the EV dream, for sure! I often reflect on the times when I ran out of battery mid-day or had to hunt for a 120V plug just so I could get enough energy for my daily commute from school. During that period, we had to go to the parking lot to hunt for an outlet or some source of power, so she could commute to her job on full

electric. This was before we owned the LEAF and before LEAFs were even for sale. Those were the days! We EV pioneers were out there blazing a trail and telling everyone we knew about a lifestyle that didn't involve gasoline or gas stations. How many worldwide conflicts are caused over oil? What a better world we would be to not rely on the limited resource of oil.

Household Energy - Of course, there are a number of ways to easily save energy. One is using a water heater blanket that goes around the heater to retain heat and installing a timer, so your water heater isn't just running hot all day while you are gone. Washing laundry with cold water is another. You can run heat and air conditioning less by dressing differently and adjusting the programming for when you are not home, and turn off lights when they are not in use. Check for rebates on new and more efficient heating and cooling systems, appliances, and smart thermostats. If you can't afford them yet, consider cheaper investments such as LED lighting and smaller appliances like fans, space heaters, toaster ovens, and pressure cookers which save you money because they use less electricity. Add inexpensive controls such as sensors, timers, and dimmers, which automate on and off and reduce energy consumption.

Figure 218 - Our home with our Volt, Roadster, and LEAF, along with my bicycle.

Now let's go beyond how we can conserve and store energy, and discuss how we can generate it, as well. Of all the renewable options out there, solar has a special place in my heart, probably because it is the most available at a household level compared to wind, hydro or geothermal. I was fortunate to have two passionate neighbors who geeked out about their photovoltaic panels at home. As we were saving up to purchase a system of our own, I did a lot of product research and ultimately spec'd out what I wanted. Unlike how most people these days opt for solar leases or a proposal by an installer, my neighbors and I decided to ask for bids on a proposal we wrote and then leveraged our buying power as a multi-household deal. Our efforts were featured in the local newspaper as we went door to door in our neighborhood to raise awareness about solar power and invited each to participate in the neighborhood discount by buying collectively.

We went with Sharp solar panels and the German-made SMA inverters. This combination represented the best value for the performance. Initially, we installed 5 kW of panels and a 7 kW inverter, 5 kW being the limit for our local utility rebate, which covered 60% of the cost. We chose a local company, owned and operated by two passionate engineers who quit their corporate jobs to brave the new industry, with a licensed roofer on staff. Many solar projects gone wrong were due to failure in structural penetration of the roof deck. We purchased a larger inverter than the PV panels we could afford to have installed at the time because it allowed us to take advantage of the rebates and provided the capacity for more panels as we save up for them. About two years after the initial install, we were the lucky recipients of some leftover stick-and-peel solar panels from a long-ago construction project of a friend's, so we built a DIY gazebo using the panels and connected them to the inverter.

Solar is magical; it is how nature builds life. There is no other source of energy on this planet that is as pure or renewable as solar. Every other form of energy in our solar system, at some point, comes from solar. Solar doesn't degrade or diminish when collected and requires a minimal manipulation of natural surroundings. It is not as detrimental to our eco-system compared to hydro and wind. The payback is not as attractive or as quick as some of the other energy conservation measures and, while it requires a significant investment to purchase, it is typically warrantied for 25 years. Studies have shown the expected output is still >80% at the end of the warranty. Solar is low maintenance, with no moving parts, and fossil-fuel-based energy is only going to get more expensive with time. For larger households, a solar hot water heater may be a less expensive option to consider. For us though, our low water consumption, paired with water conservation strategies discussed earlier, made photovoltaic a more practical investment. We are in the process of installing a 5 kW system with a SolarEdge StorEdge 7.6 kW inverter, and an Element 3 Energy 12 kWh battery. This setup means we can completely abate all energy consumption during peak hours, even on cloudy days. Batteries can store the excess solar energy generated and be recharged on the grid overnight during off-peak rates.

Evolving Our Diet - The biggest impact we can have on our lives and the health of our planet is in our diet. When I first transitioned to a plant-based diet, I did it because of the positive influence of the wonderful person who I was getting to know and learn from. I made this transition 11 years ago, and I've seen the world change around me.

Figure 219 - My wife and I have moved to a vegan diet.

Adopting a plant-based diet 11 years ago was challenging; how could I still enjoy my food the way I used to? How do I get enough protein? What do I do about pizza and the critical ingredient of cheese? As previously stated, I'm a Wisconsinite. I can't overstate just how important cheese was to me as a diet staple. There is a good reason the Green Bay Packers official fan following is dubbed "Cheese Heads." We simply adore cheese in Wisconsin, and I was no exception. At the time I transitioned, the alternatives to dairy were lacking in taste, texture and availability. Nowadays, this isn't the case; I've watched more and more products successfully infiltrate the market and gain a foothold. Examples include Silk Soymilk, Gardein plant-based "meats," a variety of excellent cheeses, Just Mayo and even substitutes for conventional eggs. These products have become available in most grocery stores, including the big-box stores like Target and Wal-Mart, across the country. Restaurants are increasingly more vegan-friendly, many with separate plant-based menus or simple substitutions they can make to their recipes to convert any meal into a plant-based experience.

At first, I didn't fully comprehend what it meant, at the global level, to make this dietary change. But through the years, I've learned quite a lot, things you don't really learn until you are doing it (living plant-based). It has become increasingly obvious to me that, in terms of environmental benefit, this is the single most positive change a person can make for the planet. In recent years, many studies have come out supporting this claim, such as the following statement from the United Nations Environmental Program (UNEP):

> "Impacts from agriculture are expected to increase substantially due to population growth increasing consumption of animal products. Unlike fossil fuels, it is difficult to look for alternatives: people must eat. A substantial reduction of impacts would only be possible with a substantial worldwide diet change, away from animal products."

As reported on www.smithsonianmag.com in 2012:

> "A United Nations report warns that 'livestock's contribution to environmental problems is on a massive scale' and that the matter "needs to be addressed with urgency,' and a report from the Worldwatch Institute[23] says that '…the human appetite for animal flesh is a driving force behind virtually every major category of environmental damage now threatening the human future…'"

According to the www.ourworldindate.org website:

> "Livestock takes up nearly 80% of global agricultural land yet produces less than 20% of the world's supply of calories. This means that *what* we eat is more important than *how much* we eat in determining the amount of land required to produce our food. As we get richer, our diets tend to diversify and per-capita meat consumption rises; economic development unfortunately exerts an increasing impact on land resources."

From the perspective of raw mass, eating a plant-based diet creates a much smaller burden on the environment. But what about water consumption? The numbers are even more staggering! The USGS states that "Producing one pound of beef requires 1,799 gallons of water.[24]" Every gram of protein in beef requires 29.6 gallons of water to create. Soybeans, for comparison's sake, require 5.8 gallons per gram of protein.

If we approach the subject as a simple matter of function, it is easy to see how much more efficient it is to consume lower on the food chain. Our individual burden on the planet's resources is far less. This is one of the strongest arguments that can be made for adopting a plant-based diet and is one that I identify with strongly at this point in my life. The wonderful thing is, this trend is growing in America, and this growth is making it easier and easier to adopt a plant-based lifestyle. It doesn't require us to give up the delicious flavors and addictive richness of our animal-based foods because many excellent substitutes have been developed and are readily available.

I will reiterate: following a plant-based diet is the single most positive-impact decision I've made, and this choice is available to anyone. Eating fewer animal products is obviously beneficial, but the most significant results require a fully plant-based diet.

Being born and raised in Wisconsin, I grew up a meat and dairy loving person. I used to eat exorbitant amounts of cheesy meat-based products and drank copious amounts of milk at nearly every meal. From the perspective of primal desire, it was not easy to transition to a plant-based diet. My first decision was to move from my Standard American Diet (SAD) to a 6-days-a-week vegetarian diet. I designated one "cheat" day every week and would consume meat products on that day only. After six months, I knocked out the carnivorous day and became fully vegetarian. As a vegetarian, I continued to consume dairy at my typical levels.

After 2 years living as a vegetarian, I was ready to go totally plant-based. The most difficult thing for me to give up was dairy, especially cheese, and especially the melted deliciousness that is a fine pizza covered with the stuff! I found myself every so often sneaking in a bite or two of a "real" cheese pizza to fulfill "pregnant-lady—level" cravings! Eventually, I found more strength in myself and was able to ignore these cravings until they no longer presented a serious challenge for me to stay true to my dietary choice to be fully plant-based. Although every so often I may cheat here or there and consume something that I know has dairy in it, at this point, after a journey of 12 years, I find it is not challenging at all to maintain a plant-based diet. Queenie and I have collectively eliminated the environmental demand of approximately 28 years of Standard American Diet consumption. This adds up to a HUGE positive impact on the planet and is a major part of our race to net-zero!

Community and Activism

Being active in our community is so important. It gives us a chance to get to know like-minded people and share a great support system. I have been very fortunate to be a part of a very caring community, both locally and online. Joining a local electric car club allowed me to meet other passionate EV fans, see their vehicles, hear first-hand insights about any EV news and development, and discuss the joy of a net-zero lifestyle. Some EV groups host and participate in periodic Cars and Coffee Events--a car and driver "meet and greet" at a designated parking lot which often hosts some real head-turning automobiles. I met some of the nicest, most eco-friendly people through our local club. They have helped me personally and professionally. We help each other with car repairs, upgrades, troubleshooting, advocacy and more. If you are in the market for a new vehicle, your local EV club is an invaluable resource.

Online, I am very active on the Tesla Motors Club Forum. I have met some successful leaders through the forum. Full disclosure: Just like anywhere else, Tesla owners are a diverse group. For example, I recently reached out to see if anyone would be willing to bring their vehicle to a school where my brother teaches for a show and tell with his students. I got mixed reactions from owners which ranged from being outraged at my audacity to ask for such an expensive favor, to owners who not only agreed to lend me their vehicle but also defended me. I have been able to connect with other owners to trade parts and bounce around ideas.

There are other online forums and groups via social media like Facebook and on apps like PlugShare, where you can leave comments about the status of a charging station and be connected to a community nationwide. This is especially helpful when traveling with an EV. My wife often checks whether a specific DC Fast Charger is working because these units are heavily used, tend to fail more frequently, and are fewer in number. This makes knowing their status more important if one is relying on them.

In Conclusion

It has become painfully clear that we humans as a species, ARE doing harm to our environment and the Earth that supports us. It has never been more important to practice ahimsa, the virtue to do no harm, which my wife and I work hard to incorporate into our decisions and lifestyle. We hope you were able to glean some nuggets from our journey in order to make your journey, to a net zero energy lifestyle easier to visualize and achieve.

Chapter 15 - The Solar Decathlon   by Ron Larson & Dr. Danica Jean Larson

Editor's note: The main author/designer/editor for this chapter was the Larsons' daughter: Dr. Danica Jean Larson, who experienced many of the events described below, and edited her father's responses to questions she thought best for this book. Ron is an early pioneer and policy setter for both solar and electric vehicles.

Quick Facts:
> **Net Energy Emissions** – 0 lbs. - 100% less than the typical American family
> **Home** – Solar Decathlon base home (670 sq. ft.), expanded to 2,200 sq. ft. in Golden CO, USA
> **Technologies Employed** – Passive solar, solar PV, 10,000-gallon thermal water, electric vehicle
> **Annual Energy Costs** – $0, down 100%

The Larson home began as a University of Colorado at Boulder, Solar Decathlon entry in 2002. The home is shown on the national Mall in Washington DC, where the team won the competition.

But we have gotten ahead of ourselves as Ron's interest in the environment and clean transportation began over 30 years earlier when he was a professor at Georgia Tech.

Figure 220 – The Larson home began as the UC Bolder, Solar Decathlon entry. The team claimed 1st place.

Danica: Your solar home has a great backstory of student competitions. But I think you initially got involved because of an interest in electric cars?

Ron: Yes, because of being a faculty advisor--first for electric cars, and much later for solar. Student competitions are a big part of why we are talking about this house and cars in this book.

Danica: You initially helped students in a car race at Georgia Tech?

Ron: This was a national race from Boston to Pasadena, held in 1970. It was called the Clean Air Car Race. I was a professor of Electrical Engineering at Georgia Tech, and my students wanted to enter an electric car in that cross-country race. Unfortunately, the "Ramblin Reck" car had to be towed from Indiana to the finish line. A photo from Life Magazine shows students ahead of the industry, almost fifty years ago.

Figure 221 - How a bunch of college kids convinced Detroit to cut smog in 1970.

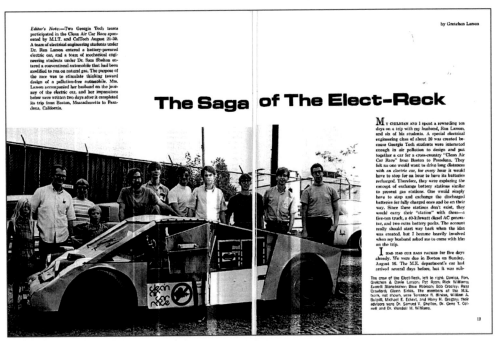

Figure 222 - Our Ramblin Reck entry from Georgia Tech.

My wife, Gretchen Larson, wrote an article for the magazine about our entry. We are on the left in the picture. We were all hooked by that first experience and continued for two more car competitions in Detroit. These were called UVDC (Urban Vehicle Design Competition). The first entry was all metal, built on a VW frame and carried 12 large lead-acid battery packs (recharged on the move in a companion van with a diesel generator). The later competitions were with a car with a fiberglass body and rechargeable battery.

These contests piqued my interest in both alternative energy and electric vehicles.

We bought our first electric vehicles--actually more of a golf cart--for you (Danica) to drive to school when you first got your driver's license in 1975. It went a maximum of 45 miles per hour--so no hot-rodding for my new teen driver.

Danica: So how did you get involved in solar?

Ron: For me solar began in 1973; I took a sabbatical from Georgia Tech to Washington DC as one of nine first-ever Congressional Fellows representing EEs. I went hoping to work for a brand new, not yet opened, congressional office - the Office of Technology Assessment. The OTA project I monitored was on solar total energy, which we now have in this chapter's house which combines solar electric and solar thermal. Before OTA opened, I worked for the House Science Committee on two bills. The first was for solar residential demonstrations, the second establishing what is now the National Renewable Energy Laboratory (NREL).

Figure 223 - Solar for me began in 1973, and I'm still using it today.

Danica: That must have been an exciting time--so much new research and excitement about solar, and support from the administration.

Ron: It was. The first Arab Oil Embargo also happened in 1973. There were very few votes against the committee's two bills or any of the OTA work. Although exciting for many Congressional staffers, I believe I was the only Congressional staffer working full time on renewable energy. I also served for a time on a Joint Small Business Committee, headed by Senator Gaylord Nelson, on a huge solar report. The Federal renewable energy budget grew from about $1 million to $13 million in my two years in D.C.

Danica: Did you teach alternative energy when you returned to Georgia Tech?

Ron: Even earlier, in 1972, our Electrical Engineering department focused on a national competition called SCORE (Student Competitions on Renewable Energy), that grew out of the first student car competition. In 1976, after my return, Georgia Tech had several teams, with one winning the solar electric competition using a pre-1900 relic steam engine and 800 square feet of reflector.

Danica: So, all that experience is what got you the job in the newly created SERI (Solar Energy Research Institute) and spurred the move to Denver?

Ron: Yes, I moved to be part of the Solar Energy Research Institute, which later became part of the National Renewable Energy Laboratory (NREL). That job had a policy orientation, consistent with my Congressional renewable energy policy experiences. I was part of the formation of the organization and so was in Golden when it opened on July 5, 1977. It recently celebrated its 40th anniversary.

Danica: You were involved in the first Earth Day celebrated in Colorado?

Ron: Yes, for Colorado in 1990, but only a little for the first Earth Day in 1970. At that time my interests at Georgia Tech were all still in Electromagnetics (Maxwell's Equations), and just heading towards electric cars. But I later worked at SERI for Denis Hayes, then and still Earth Day's main coordinator. He is now a friend and a renewable energy authority all readers of this chapter should pay attention to. He so greatly influenced my life that I later helped form "Golden Earth Days" in 1990, which eventually was the main organizer of the annual Colorado Solar Home Tour. Our home, along with that of Steve Stevens (chapter 5), was on that tour three times--once while still under construction in 2004.

Danica: Then you got involved with a new student competition at CU Boulder.

Ron: That competition, still ongoing, is called the Solar Decathlon, where college students design, build and display a solar home. The competition began under Department of Energy sponsorship starting in 2002 on the mall in Washington DC. Because of the student car and solar competitions I've described above, I had a great sympathy for Professor Mike Brandemuehl, the CU-Boulder faculty advisor of the team.

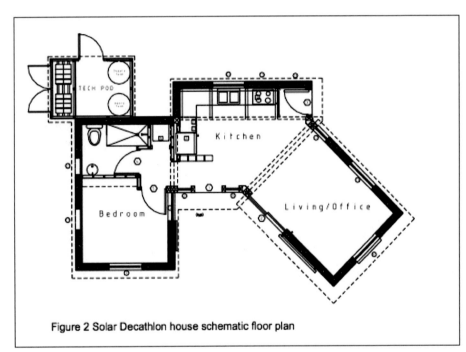

Figure 2 Solar Decathlon house schematic floor plan

Figure 224 - Solar Decathlon house designed by Bolder students. I agreed to buy this house at the end of the competition.

Danica: The competition at CU-Boulder ran into some trouble, not due to lack of student interest, but funding?

Ron: Yes. In early 2002, the team was nearing a point of insufficient funds. We hated to see all the students' work lost, so to fund the completion and travel costs, my wife and I agreed to buy the CU house after it was returned from D.C.

Danica: The team carried the entire tiny house off on a semi to Washington D.C. How did the competition turn out?

Ron: The students first had to de-construct the house into seven pieces and place them on three large semis. The trip took more than three full days, with numerous diversions to avoid low bridges, etc. But then the CU-Boulder team won first place! They say that one main reason was the high scores they ran up in the electric car part of the 10-part competition; they did a good job of maximizing energy into the car batteries. Their roof-size-limited solar panels were top-of-the line for efficiency, and they are still working well 16 years later. The students told us that 100,000 visitors came through the house during their period on the Mall. The Solar Decathlon is still

Figure 225 - Proudly displaying the winning entry by Colorado University students in Boulder.

active with colleges from around the world, as they continue to push the boundaries of energy efficiency.

Danica: You'd promised to buy the house when it was finished, but now, as a winner, it was more valuable. What happened next?

Ron: We found the state of Colorado could only move very slowly. After many months, they held a closed-bid auction. We increased our previous promise a bit and beat out two other bidders. But even then, the release took many more months as, once again, thousands toured the house as it sat near the CU-Boulder football stadium. The house even hosted some meetings.

Danica: You had a solar home at that time. But the property where you lived on Lookout Mountain was surrounded by trees, with a great view of Denver but no good southern exposure to expand beyond the solar hot water you had added in 1978. How did you handle that?

Ron: After viewing Denver from our A-frame home location, a 2000-foot-high mountain vantage point for more than a quarter-century, we had an opportunity to move almost anywhere. But we really liked our Lookout Mountain neighborhood and were eventually able to get just about the only remaining lot - only a few hundred yards away. This time, we faced south on an almost treeless lot, with the main challenge being a quite steep slope.

Danica: You had some challenges with a decision about how, where, and how much to add on to the house to accommodate a family, and your wife needed a pottery studio.

Ron: Yes, many challenges, most caused by the significant slope. Adding square footage was necessary - as all the Decathlon houses had a mandated maximum size of 670 sq. ft., much less than we wanted. Even the

Figure 226 - The expanded CU of Boulder Decathlon home with about triple the square footage.

CU-house's one bedroom was not big enough for a double bed. And, by now, we had a formidable collection of native arts. Gretchen was able to finally get a respectable pottery studio that we designed for later conversion into a master bedroom. And we finally could have a garage.

Danica: The remodeled house is unusual in many ways. You needed to make many decisions to support this larger square footage?

Ron: Yes, most of the alterations were again dictated by the slope and the unusual pie-shaped lot between two streets. After receiving permission from neighbors to amend the local building code, we changed the street address to make for easier access to a garage and less snow shoveling. But the overall house is now a fairly "standard" solar house. That is, most of the high-R-value windows face south, oriented in the east-west direction, and overhangs maximize winter but not summer sun. The extension walls are thicker (10 inches) and of higher R-value than even the CU house. In retrospect, I wish we had gone to at least a 12-inch wall, for even lower energy consumption.

Danica: Does the now-expanded house have adequate energy production for your heating needs?

Ron: Yes. The roof area limits the number of solar panels, but it is adequate. In large part, this is because much of the winter heating needs are met by a 10,000-gallon hot water storage tank that captures energy all summer long. This tank could be justified--in large part again--by the slope of the lot, as it is buried on the north side of the home. If we did not have the tank, much more fill dirt would have had to be moved in. The tank is composed of 8"-thick poured concrete. The tank interior has six inches of foam insulation and is lined with a thick rubber fabric. The inside dimensions are about 60 ft. by 5 ft. by 8 ft. There are two compartments of about 1000 and 9000 gallons, and we can adjust the temperature of each sub-tank separately. The water in the tank is heated from thermal solar panels on the roof and generates enough heat in the summer and fall to nearly make it through the winter months.

Danica: How hot can the water get?

Ron: The water temperature coming off the roof can be close to boiling, so I sometimes intentionally run the system at night, to cool things off. The temperature in the thermal tank at the end of summer is 160 degrees. I believe it will last all winter when I get around to investing in movable insulation. Backup heat is provided by two wood-burning stoves and a few electric radiant heaters, which are not needed very often. The south side of the house, photographed here showing the expanded house, gives no indication of this hot water storage on the house's north side. The CU house was electric only, with an air-to-air heat pump. We switched to radiant floor heating, with which we have been satisfied.

Figure 227 - South side of the expanded house on our lot above Golden CO.

Danica: Did the house's photovoltaic energy production inform your decision to get hybrid and electric cars?

Ron: No. Until recently, we have had hybrids: an early version of the Honda Civic with no rechargeability and newer Chevy Volt, which does have a plug-in battery. We would have made these choices independent of the CU home's 7 kW photovoltaics.

Danica: Anything about the location that influenced the design?

Figure 228 - Pikes Peak is visible from the roof on clear days.

Ron: Yes. From a few spots we could see Pike's Peak, which is about 50 miles away. I am standing on a part of the added portion, looking out over the original CU-house panels. Behind me is a small unheated greenhouse.

Danica: Merging the old and new must have been quite a challenge for the architect?

Ron: Yes, the architect, Walt Kaesler, did a great job getting past all the county permit issues. Working with Walt, Gretchen and I were able to make the key decisions, such as where to place the garage and pottery studio, now below the garage at the east end. The builder, Doug Larson, who is no relation, was great to work with, along with his small crew of first-rate carpenters. The main new architectural feature is the two-story stairway, shown at right. As an artist, Gretchen has collected folk art from around the world. She wanted a way to display our treasures. We designed this stairway together; I've yet to see one like it in any other home. It has extensive, no-cost display areas on both sides and wonderful rigidity, while still being quite open. The stairway pieces are bolted together, with each vertical carrying the load of eight stairs.

Figure 229 - Our custom stair design, used to display artwork

Danica: You have a Tesla now. Was that a difficult decision to invest in such an expensive car?

Ron: Yes and no. We were strongly influenced by the Tesla ownership (and sales pitch) of two of our closest friends. We felt it important to support Elon Musk and the Tesla company in late 2016, with the all-electric non-hybrid we had always wanted and know is needed for the environment. Our Model S has been used for camping, including three multi-day trips in the past year, and we are delighted to have a free electricity stop, every few hundred miles. The car's acceleration and safety systems are a joy; I've never felt safer in a car. But, of course, we wish it had been less expensive.

Figure 230 - Facing south, by our Tesla Model S, powered by our solar array.

Danica: You finally retired, but have kept working--maybe more hours than when employed?

Ron: Yes, after retirement more than 20 years ago, I was a co-founder of CRES, the Colorado Renewable Energy Society, and later a term as Chair of ASES, American Solar Energy Society, where I was proud to be elected as a Fellow. I'm sure such continued activity is important for all the authors in this book because it is important to convince others of the urgent climate need.

Danica: Your newest passion is, in reality, an ancient technology (Terra Preta) newly dubbed "Biochar?"

Ron: Yes, I have switched my attention away from the now common and increasingly cost-effective solar and wind policy topics. Climate policy issues are now my focus, not the issue of running out of fossil fuels, which was my passion for years. When we started on this house, the term "biochar" had not even been invented. However, I was working on improved charcoal-making cooks stoves--having led a USAID project in Sudan. Biochar seems to be the best way we have for taking carbon out of the atmosphere. One basically heats wood to cook off the hydrogen and water, leaving carbon, which is a great amendment for soils. I am on the board of the United States Biochar Initiative. USBI is a daughter of the International Biochar Initiative, where I am on its technical advisory committee. See www.biochar-international.org. I am trying to mainly stick with international and policy aspects of biochar, in the half of Geoengineering called Carbon Dioxide Removal (CDR). This is a small climate-solution area in between mitigation (carbon neutrality) and adaptation.

Danica: You have encouraged other solar homeowners to share their enthusiasm for solar by initiating the first Colorado Solar Home tour?

Ron: Yes. I did that shortly after I not very successfully retired. Now the solar home tour has grown into a major annual Colorado event, always coupled with the national event sponsored by the American Solar Energy Society, which takes place the first weekend in October. Readers will get a lot out of such a tour, happening in almost every state. If folks would like to see more, Jim Smith did a nice 15-minute video tour that included our home: https://www.youtube.com/watch?v=zgR-lI8eQKE

Danica: What else should folks interested in pursuing solar energy for their current or future home gain from your experience?

Ron: First, going solar will be a great learning experience. It will take a lot of time. But there are now many available resources, of which this book is a great example. Emphasize passive solar and energy efficiency. Do not skimp on the energy efficiency measures--think thick walls. It won't all be fun, but you will be glad you did it--for your grandchildren. We have five, all grand-daughters, shown here. I'll bet that most of the authors in this book were motivated by concern for their grandchildren. So, this final photo is a good way to end this Larson daughter-father solar/EV chapter.

Figure 231 - Our five granddaughters, the real reason for our journey.

Summary

We hope you enjoyed these stories from real people drastically reducing their carbon footprint. If everyone took such journeys, we could, in short order, take the initial necessary steps to solve the global-warming problem. Rather than its being a sacrifice, we hope you see that in many ways, life is better. Families living in zero net energy homes are self-sufficient and feel liberated from using dirty fossil fuels. Rather than being prohibitively expensive, for about the cost of a kitchen remodel[25] one could be energy independent. If by chance you need to move early, you should recoup your investment in solar with a better resale price on your home, much better than that kitchen remodel.[26] And do not forget the considerable monthly savings in energy costs. Savings range from $225 to $750 per month for the home and transportation, with several of the families now making money (chapters 2, 5 & 7). This is a savings--after taxes and indexed for inflation, a solid return for everyone featured in this book.

While this book focused on energy emissions, we emphasize that one's diet and purchasing habits can also greatly affect one's carbon footprint. We mostly glossed over this, not because it is not important, but because hard data is very difficult to come by. As representative examples, several of the stories (Chapters 1, 4, 5, 8, 9, 12, & 14) feature gardening and a low--or no--meat diet as part of their journey.

While there are several paths one could take to become more energy efficient, there are some common themes we want to highlight.

A journey, not a big bang – For nearly all the homeowners the journey to net zero energy use took years, sometimes decades, not a one-time big bang. Part of this is due to the reality that many of the necessary technologies were not available when we began our journeys. But a big part of this is learning and applying lessons learned as we go. Most of us measure, test and adjust, and without thought, we are making mid-course corrections to our trajectory. We find a leaking door or window and make it energy tight. We discover a hole in insulation and plug that hole. So, don't use the excuse that you can't do it all as an excuse for doing nothing, because even the small initial steps bring results and satisfaction.

Figure 232 - The journey can be as enjoyable as the destination.

Insulation – Everyone paid close attention to insulation and air infiltration to make their homes as energy efficient as possible. Normally, it is far more cost-effective to reduce one's need for power than to add more solar panels. While it is easier to incorporate passive solar and/or PassivHaus technology into new construction, existing homes can be upgraded as shown in chapters 1, 5, 8, & 10. For those wanting a traditional home look, Mark and Lloyd, in chapters 2 and 11, showed conventional homes built with energy saving technologies with solar incorporated in outbuildings or ground mounted.

Transportation – Cars use a lot of energy, and there is no way to get to zero net energy without moving to plug-in technology coupled with solar for charging, or a strong reliance on public transportation and bicycles, as Pete and Victor do chapters 8 & 10. Families with 100% electric vehicles normally charge at home and feel empowered and free from polluting gas stations. Electric vehicles have two large advantages over internal combustion engines (ICE). The first is efficiency, as an ICE only converts about 25% of the stored energy in gasoline into forward motion. For an EV, that number is about 75%.[27] Accounting for extraction, transmission, and conversion, (called a well-to-wheels or end-to-end efficiency calculation), an EV is over 200% more efficient than gasoline[28]. Then while an ICE requires fossil fuels, an EV can be powered from *any source*, including cleaner geothermal, solar, wind, hydro or nuclear generated electricity. In this book, every featured family has made the move to at least one plug-in car, and, if they retained a gas-powered car, it was one of the more fuel-efficient ones available. While electric vehicles require some compromises, those compromises have greatly diminished in recent years. Long-range travel is now easy with a Tesla, thanks to their nationwide Supercharging network. Superchargers can add over 200 miles in 30 minutes of charging, and there are over 1340 Supercharger locations with more than 11,000 plugs.[29] Tesla has also been aggressive in building out destination chargers, which, while slower, can add a respectable 50 miles/hour at many (3600+) restaurants and hotels. With the long-awaited mass market, Tesla Model 3 shipping, one can tap into that network of fast charging for just over $35,000, very close to the average new car price of $33,000.[30] If one buys a long-range LEAF or Chevy Bolt, both the CHAdeMO and SAE Combo fast chargers are also expanding, with both numbering over 1800 locations in the USA.[31] These chargers add over 100 miles/half hour of charging. As part of the VW "diesel-gate" scandal settlement 2,800 new charging stations are being deployed across the U.S.[32] If you plan charging around mealtimes, it is easy to drive 400+ electric miles in a day or 600+ in a Tesla.

With the Tesla Model X, one can tow a camper or large boat, uphill to that mountain lake, with ease. Chrysler makes the plug-in Pacifica, giving minivan functionality with an impressive EPA MPGe of 84. The only hole now remaining in the fossil-fuel transportation sector is long-haul, heavy-freight transport. Even that should soon be addressed by several OEMs as Tesla, Volvo, and Mercedes plan to release trucks by 2019, so a consumer pickup truck cannot be far behind.

Figure 233 - Tesla to produce a Semi with up to 500 miles range pulling 80,000 lbs.

Solar – While it is possible to generate your own power with wind or hydro, rarely are those options cost effective for urban homeowners. For a wind turbine to be effective it needs to be 100' above the tree line. That requires a very tall, meaning expensive, tower and is a significant zoning issue for most people. For hydro, one needs ready access to a strong-flowing creek or river, normally

Figure 234 - Installing solar puts local craftspeople to work.

requiring significant permitting. Again, those technologies rarely work at a homeowner level. Fortunately, solar works well for many homes, and zoning is rarely an issue. Many people can add solar invisibly to their homes, like the Erb's, Norby's and Steven's in chapters 1, 4 & 5, who installed it on the backside of their roof, so nothing is visible from the street. Or consider the Bishops in chapter 2, who placed their solar on their barn/workshop, leaving their house conventional-looking or the Marcum's in chapter 11, who used ground-mount solar in an inconspicuous area of their property. Once the new Tesla solar roof starts shipping in volume, one can have a solar roof that blends into any neighborhood.

Lighting – Another common theme among all the homeowners has been the conversion to all-LED lighting. In a typical home, 10% of its power is for lighting, and LED's can reduce that amount by 85%. LED's are now quite affordable and can be found priced BELOW energy-wasting incandescent bulbs. There is no need to wait for bulbs to go out, as an LED easily pays for itself in energy savings alone. Remember, 95% of the energy consumed by the old-fashioned Edison-type incandescent bulb is spent creating heat, while only 5% of what you're paying for produces visible light! That extra heat from incandescent lights also adds to your air conditioning loads.

Summary -- All the households featured took time for their journey, used good insulating practices, and installed LED lighting to make significant reductions in emissions and energy costs. Most also employed rooftop solar and plug-in vehicles for dramatic reductions. Other techniques and technologies can and should be used when appropriate. Yes, it costs money to make these changes, but now that you understand the ramifications of conservation, the loss of potential savings will gnaw at your psyche. You know that overall, you would be wasting instead of benefiting from less-expensive energy.

Hopefully we have convinced you to begin your own sustainable journey to reduce your emissions and save some of your hard-earned money.

MIND YOUR Ps AND Qs, AND REMEMBER YOUR TLAs

Every specialty has its jargon, and sustainable transportation is no exception. Sometimes, the hardest part can be keeping track of the three-letter acronyms. In order to avoid defining them in every chapter, here's a glossary of some TLAs, a few FLAs, and other terms you may encounter, both in this book and when reading about sustainable transportation generally.

APU – Auxiliary Power Unit

> A device, such as an engine, which converts a non-electric store or source of energy into mechanical or electrical form. Note that many HEV development engineers will refer to an HEV's non-electric power unit as the APU, even (as is frequently the case) when it is the vehicle's primary means of propulsion.

Aquaponics –

> It's Aquaculture (fish farming) + Hydroponics (Growing plants without dirt). The fish waste fertilizes the plants. The plants purify the water.

BEV – Battery Electric Vehicle

> An EV which stores all of its energy in on-board batteries and derives all of its propulsion by using one or more electric motors to convert that stored energy into mechanical form. The batteries may (very rarely) be primary batteries, which are refueled by removing and replacing their spent chemical contents after discharge. Typically, though, BEVs use secondary batteries, which are recharged by plugging them into an off-board electric source.

Bitcoin –

> Bitcoin is a cryptocurrency and worldwide payment system. It is the first decentralized digital currency, as the system works without a central bank or single administrator.

CARB – California Air Resources Board

> A California state agency with broad power to set vehicle emissions standards. Because CARB's vehicle regulatory activity predates that of the U.S. Environmental Protection Agency (EPA), it is the only state agency allowed to do so, though other states can adopt CARB standards.

CFRP – Carbon Fiber Reinforced Polymer

> A lightweight but strong material used in the BMW I3 and other cars

CSHEV – Charge Sustaining Hybrid Electric Vehicle

> An HEV which receives energy only in the form of fuel and sustains battery charge through the way the APU and regenerative braking are controlled. The electric components in a CSHEV powertrain enable fuel to be burned more efficiently and with significantly lower emissions than is possible with an engine alone. The original Toyota Prius is a CSHEV, as are all other mass market passenger HEVs sold in the U.S. before the Chevy Volt was introduced in December 2010. Much of the early advertising for these cars alluded to their charge sustaining nature, with phrases like "the electric car you never plug in."

DHW – Domestic Hot Water

EV – Electric Vehicle

Though often used exclusively, to denote only BEVs, this book uses the term EV inclusively, to encompass any vehicle using one or more electric motors to provide all or part of its propulsion: BEVs, HEVs, FCVs, railcars and trolley buses receiving electricity through third rails or overhead wires, ...

EVSE – Electric Vehicle Service Equipment

External physical equipment used to connect an EV to the electric grid. An EV charging station.

FCV, or FCEV – Fuel Cell (Electric) Vehicle

A series HEV which employs a fuel cell as its APU. Fuel cells produce electricity by oxidizing their fuel without combustion. Because there is no combustion, they are cleaner than engines running on the same fuels.

GSHP – Ground Source Heat Pump

An HVAC system which uses the earth as its heat source in heating mode and as its heat sink in air conditioning mode. At depths below two meters, ground temperatures tend to be stable, at approximately the annual average air temperature, which improves the efficiency of heat transfer. GSHP systems are frequently called "geothermal" systems, though this term is more properly applied to systems that draw high temperature heat from subterranean nuclear decay, as in areas with hot springs and other near-surface volcanic activity.

HEV – Hybrid Electric Vehicle

An EV which combines one or more electric propulsion units with an APU.

HVAC – Heating, Ventilation, Air Conditioning

The unit used to heat and cool your home. Often a heat pump but could also be the furnace and/or air conditioning

ICE, ICV, or ICEV – Internal Combustion (Engine) Vehicle

A conventional vehicle, powered by an internal combustion engine (ICE), typically burning gasoline or diesel fuel. When rude ICV drivers park in EV charging spaces, blocking access to the chargers, EV drivers refer to it as being "ICE-d."

ICF – Insulating Concrete Form

A Styrofoam form that snaps together; but when filled with rebar and concrete make a very solid and well insulated home. Size is typically 16" tall, 4' long and 11 ½" in depth.

Level 1 Charger –

A 120V charging unit for electric vehicles. At 8-15 amps this is a slow charge allowing one to charge about 50 miles in a 12-hour period.

Level 2 Charger –

A 240V charging unit for electric vehicles. At 16-80 amps it allows one to fully charge large batteries overnight.

Level 3 Charger – also known as fast DC charging

A 400V fast charging unit for electric vehicles. Currently there are three standards. Tesla Superchargers at 130 kW, CHAdeMO (Nissan & Mitsubishi) mostly 50 kW, and DC Combo (Chevy, BMW & VW). Most are 50 kW, but plans are for up to 300 kW.

Mini-Split Heat Pump –

A mini-split is a small efficient heat pump normally between 9,000 – 24,000 BTU or between ¾ ton and 2 tons that forgoes normal ductwork and has just a small indoor and outdoor unit.

MJ – megajoule

A unit of energy equal to .2778 kWh or .00759 gallons of gasoline.

Parallel HEV –

An HEV in which mechanical output from the APU is conveyed to the wheels in mechanical form. Conceptually, the APU output follows a path "parallel" to the mechanical path taken by the electric motor output, even when the two paths are actually one and the same. It is important to note that series and parallel architectures involve aspects of both hardware and control strategy. The first-generation Chevy Volt, primarily a series HEV, employed hardware which enabled parallel operation in certain corners of the operating envelope. The second-generation Volt achieves much of its improved efficiency by expanding the extent of parallel operation. The Toyota Prius, like many nuclear submarines, uses hardware that is primarily parallel, but includes some series operating modes.

PEV – Plug-In Electric Vehicle

Any battery electric (BEV) or plug-in hybrid electric vehicle (PHEV).

PHEV – Plug-In Hybrid Electric Vehicle

An HEV with sufficient battery capacity to allow meaningful amounts of propulsion to be provided by depleting the battery charge while driving, then recharging from the electric grid. Some engineers refer to PHEVs as charge depleting HEVs (CDHEVs), a practice which gives marketing people migraines, due to the image it evokes of being stranded by the side of the road with a dead battery. PHEV has been the preferred terminology since about 2003.

PV – Photovoltaic

Literally, "light electric," denoting a device which converts solar energy into electricity.

RE – Renewable Energy

Any form of energy which is replenished within a human, rather than a geologic, time frame.

ROI – Return on Investment

Series HEV –

An HEV in which the APU drives a generator to convert fuel into electricity, and the wheels are driven only by electric motors. Conceptually, any fuel input to the vehicle must follow a "series" of conversions which goes through electric form in order to get to the wheels. If APU output exceeds driver demand, it is stored in batteries. If driver demand exceeds APU output, the electricity goes directly to the electric motor(s) and is augmented by energy from the batteries. In the case of series HEVs which lack batteries, such as railroad locomotives and open pit mining dump trucks, excess APU output and regenerative braking energy are burned off in resistor banks. The BMW range extending I3 is an example of a series HEV.

SIP – Structural Insulating Panel

A panel made of Styrofoam sandwiched between two pieces of plywood. This form of construction offers better insulation and lower air infiltration than typical construction.

TLA – Three letter acronyms

WWS – Wind, Water, and Solar

The three most sustainable forms of renewable energy. At their core, all originate as energy from the sun.

ZEV – Zero Emission Vehicle

A CARB regulatory construct, denoting a vehicle with no tailpipe emissions. ZEV also sometimes describes an electric-only operating mode, or a location where ICV operation is forbidden.

Endnotes

[1] http://www.ucsusa.org/clean-vehicles/electric-vehicles/life-cycle-ev-emissions#.VspxnU0UXDc

[2] https://www.climatecommunication.org/wp-content/uploads/2011/08/presidentialaction.pdf

3 https://www.epa.gov/ghgreporting/ghgrp-refineries

4 http://www.ukweatherworld.co.uk/forum/index.php?/topic/62783-dr-richard-alleys-horse-ploppies/

[5] http://data.worldbank.org/indicator/EN.ATM.CO2E.PC

[6] http://data.worldbank.org/indicator/NY.GDP.PCAP.CD

[7] https://www.google.com/fusiontables/DataSource?docid=1l4OH-lUAolL-fY9v3Ka1X7mRzUoxk5iYwBRauc0

[8] https://www.eia.gov/dnav/pet/pet_pnp_unc_dcu_nus_a.htm and https://www.epa.gov/ghgreporting/ghgrp-refineries for the refining portion.

[9] http://www.eia.gov/tools/faqs/faq.cfm?id=307&t=11 and https://www.epa.gov/ghgreporting/ghgrp-refineries for the refining portion.

[10] https://www.eia.gov/tools/faqs/faq.cfm?id=74&t=11

[11] http://cdiac.ornl.gov/pns/faq.html

[12] https://www.eia.gov/environment/emissions/co2_vol_mass.php

[13] https://www.eia.gov/environment/emissions/co2_vol_mass.php

[14] http://clevelandhistorical.org/items/show/63#.VoQZAU1gnDc

[15] Michael Rotman, "Lake Erie ," *Cleveland Historical*, accessed December 30, 2015, http://clevelandhistorical.org/items/show/58.

[16] http://www.forbes.com/2010/04/29/cities-livable-pittsburgh-lifestyle-real-estate-top-ten-jobs-crime-income.html?boxes=Homepagetoprated

[17] http://www.ucsusa.org/clean-vehicles/electric-vehicles/life-cycle-ev-emissions#.VspxnU0UXDc

[18] www.sunplans.com

[19] www.sunplans.com

[20] Different state/power companies have different methods of treating energy credits. We are fortunate that our state/power company offers a 1 to 1 kWh credit arrangement. There is no annual truing up so credits can accumulate indefinitely. This could change for us in the future as they are considering other programs.

[21] https://www.bullfrogpower.com/

[22] https://www.quora.com/How-much-coal-is-required-to-generate-1-MWH-of-electricity

[23] http://www.worldwatch.org/node/549

[24] https://water.usgs.gov/edu/activity-watercontent.html

[25] https://www.homeadvisor.com/cost/kitchens/remodel-a-kitchen/

[26] https://www.forbes.com/sites/ashleaebeling/2011/08/01/how-much-do-solar-panels-boost-home-sale-prices/#3e3cc9ae11f2

[27] https://matter2energy.wordpress.com/2013/02/22/wells-to-wheels-electric-car-efficiency/

[28] https://www.linkedin.com/pulse/well-wheel-why-electric-vehicles-evs-much-more-than-alade-kolawole/

[29] https://www.tesla.com/supercharger

[30] https://www.usatoday.com/story/money/cars/2015/05/04/new-car-transaction-price-3-kbb-kelley-blue-book/26690191/

[31] https://www.afdc.energy.gov/fuels/electricity_locations.html

[32] https://jalopnik.com/heres-where-vw-plans-to-install-electric-vehicle-fast-c-1825775383

Made in the USA
Lexington, KY
29 October 2018